VEGETATION
OF THE
PEAK DISTRICT

VEGETATION

OF THE

PEAK DISTRICT

by

C. E. MOSS

B.A. (Cantab.), D.Sc. (Vict.), F.R.G.S., F.L.S.,

Curator of the Herbarium, University of Cambridge

Cambridge:

at the University Press

1913

CAMBRIDGE
UNIVERSITY PRESS

University Printing House, Cambridge CB2 8BS, United Kingdom

Published in the United States of America by Cambridge University Press, New York

Cambridge University Press is part of the University of Cambridge.

It furthers the University's mission by disseminating knowledge in the pursuit of education, learning and research at the highest international levels of excellence.

www.cambridge.org
Information on this title: www.cambridge.org/9781107688131

First published 1913
First paperback edition 2014

A catalogue record for this publication is available from the British Library

ISBN 978-1-107-68813-1 Paperback

PREFACE

THE study of vegetation in the British Isles, begun by the late Robert Smith, is being vigorously prosecuted by the members of the British Vegetation Committee. Already, several vegetation maps and memoirs have been published of parts of the central and northern Pennines, Scotland, Ireland, and Somerset by W. G. Smith, Lewis, Pethybridge, Praeger, Rankin, and myself, in addition to several minor publications by these and other members. Whilst this book was going through the press, Tansley's *Types of British Vegetation* appeared, where, for the first time, a sketch of the plant formations and plant associations of the whole of the British Isles is given. Several vegetation maps, of Hampshire, the Isle of Wight, Norfolk, north-eastern Yorkshire, Lanarkshire, and other districts, have been finished by various members of the Vegetation Committee, but cannot be published at present owing to lack of funds. The present volume and the accompanying maps owe their publication to the generosity of the Royal Society and the Royal Geographical Society, whom I take the present opportunity of thanking on my own behalf and on that of British phytogeographers and ecologists in general. I fear, however, that, until government recognition is taken of the botanical survey of the country, publication of this kind of work will continue to languish.

The present work is the result of a botanical survey of the Peak District of the southern Pennines begun in January, 1903. In preparing the vegetation maps, the Ordnance maps on the scale of six inches to the mile (1 : 10560) were used for field work. However, these were not found so superior to the one-inch maps as had been anticipated, owing to the fact that the six-inch maps of the moorlands of the district, with the exception of those in the West Riding of Yorkshire, are not contoured.

With regard to the nomenclature of plant communities, the terms plant formation and plant association are used in accordance with resolutions passed unanimously by the British Vegetation Committee, and presented to the International Congress of Botanists held at Brussels in 1910. They are used in the same sense throughout Tansley's *Types of British Vegetation.*

The names of plants are, as a rule, the same as those given in the tenth, the latest edition of *The London Catalogue of British Plants* (London, 1908). This being so, the author-citation is omitted, as being unnecessary in a work of this character: synonyms, however, are added in special cases. The sequence adopted is that of Engler's system which, in several European countries and in the United States of America, is rapidly superseding that of Bentham and Hooker.

I wish to thank Mr J. Ramsbottom, B.A., of the British Museum (Natural History), for kindly reading the proof-sheets, the Royal Geographical Society for use of the blocks of figures 4, 12, 15, 22, 24 and 25, and Mr A. Wilson, F.L.S. for use of the blocks of figures 19, 30 and 31.

C. E. M.

CAMBRIDGE,
December 1912.

CONTENTS

CHAPTER I

INTRODUCTION

CHAPTER II

WOODLAND ASSOCIATIONS

CHAPTER III

SCRUB ASSOCIATIONS

CHAPTER IV

GRASSLAND ASSOCIATIONS

CHAPTER V

ASSOCIATIONS OF ROCKS AND SCREES

CHAPTER VI

MARSH AND AQUATIC ASSOCIATIONS

CHAPTER VII

MOORLAND ASSOCIATIONS

CHAPTER VIII

CULTIVATED LAND: CULTURE ASSOCIATIONS

APPENDIX I

SUMMARY AND RELATIONS OF THE PLANT COMMUNITIES OF THE PEAK DISTRICT

APPENDIX II

SUMMARY OF BRITISH PLANT FORMATIONS AND ASSOCIATIONS

APPENDIX III

LIST OF ILLUSTRATIONS

*available for download in colour from www.cambridge.org/9781107688131

CHAPTER I

INTRODUCTION

General description of the Peak District. Types of scenery. Rocks and soils. Soils and their characteristic plants. Flora and vegetation. Vegetation maps. Plant communities. Vegetation maps and floristic maps. The value of vegetation maps. Rainfall. Smoke. Temperature. The upper atmosphere ; temperatures ; direction of the wind ; velocity of the wind ; humidity of the atmosphere. Note on the use of the words "acidic" and "basic."

GENERAL DESCRIPTION OF THE DISTRICT

THE Peak District has no definite geographical boundaries; and, for the purposes of the present memoir, it is regarded as being co-extensive with the accompanying vegetation maps (see also figures 1 and 2). A large proportion of the district consists of unenclosed moorland and grassland; and there are numerous small vestiges of scrub and primitive woodland, besides several comparatively extensive stretches of semi-primitive woodland. Cultivated land ascends the valleys, usually up to about 1000 feet (305 m.), and occurs also as more or less isolated "intakes" up to about 1500 feet (457 m.). Most of the cultivated land is utilized as permanent pasture; and there is very little arable land. Plantations are fairly numerous; and a few of them are of moderately large size.

The highest elevation of the district is reached on an extensive, undulating plateau which bears the singularly inappropriate name of "the Peak." This plateau, the highest in England south of the mid-Pennines, is peat-clad; and it attains an altitude of 2088 feet (636 m.). North of the Peak are two summits which attain heights of more than 2000 feet (610 m.): one of these, known as Bleaklow Hill, is situated six miles

(9·7 km.) to the north of the Peak, and is 2039 feet (621 m.) in height; and the second, one mile south of Bleaklow Hill, and apparently without any special name, is 2068 feet (630 m.) high. These three are the only Pennine summits, south of the "Yorkshire giants" of Whernside (736 m.), Ingleborough (723 m.), and Pen-y-ghent (686 m.), which reach a height of 2000 feet.

The westerly slopes of the Pennines descend rather abruptly into the lowland plain of Lancashire and Cheshire; and this physiographical feature is reflected on the accompanying vegetation maps by a comparatively narrow western zone of heather moorland. On the east, the slope is more gradual; and the zone of heather moor is correspondingly wider. The higher plateaux are covered by cotton-grass moors and bilberry moors, and the steeper hill slopes by uncultivated grassland. The foot of the western slopes of the Peak District is characterized by a densely populated manufacturing district, of which Manchester is the centre; and Sheffield is the centre of another manufacturing district which lies at the foot of the eastern slopes. The Pennine moors stretch away from the Peak in a northerly direction; and it is almost but not quite possible to walk along the Pennine watershed from the Peak to the Border without leaving the uncultivated land. South of the Peak lies the sequestered valley of Edale; and to the south of this the limestone hills and dales are situated. The limestone area is flanked, both on the east and on the west, by a southern extension of characteristic Pennine moorland and grassland. The lowest altitudes occur where the streams leave the district, usually at an altitude of about 100 metres. The streams harbour a few aquatic plants; but the latter are, on the whole, poorly represented on the Pennines generally.

In pre-railway days, the Pennine hills, with their peat-clad, unfenced, and undrained summits formed an effectual barrier between the Lancashire and Yorkshire peoples. Before the construction of the turnpike roads, about a century ago, the Pennines could scarcely be crossed except by the primitive pack-horse roads. Some of these still exist as public footpaths; but others, it would appear, have been closed, and are now largely overgrown with rough grasses. At the present time, the southern Pennines may be crossed by half a dozen good roads, two canals, and four or five railways. An interesting account of a journey across the Pennines in the early part of the eighteenth century is given by Defoe (1725: 90, *et seq.*)[1].

Portions of five counties, namely, south-east Lancashire, north-east Cheshire, north Staffordshire, north Derbyshire, and south-west Yorkshire, are represented in this district; and on the high moorlands several of the head-streams of the Mersey, Dee, Trent, and Yorkshire Ouse take their rise.

Types of Scenery

The district furnishes some interesting and distinct types of scenery, which depend primarily on the nature of the geological strata (cf. figures 1 and 2).

The lower hills of the north-west and north-east of the district are composed of sandstones and shales belonging to the Coal-measure series. It is on or near these rocks that the manufacturing areas are situated. The hills of the Coal-measures are usually cultivated up to their summits. Arable land, whilst nowhere really common, is more abundant on the Coal-measures than elsewhere; and more wheat is grown on such soils than on any other soils of the district. This is an interesting fact, as the soils of the Coal-measures are usually described as cold and backward (*e.g.*, by Lees, 1888: 66). The uncultivated parts of the Coal-measures are few and often isolated, and consist usually of heather associations on the sandstones and of grassland associations on the shales. However, on the few areas of uncultivated land of the Coal-measures at the higher altitudes, heather moors and cotton-grass moors occur, as, for example, south-west of Buxton. The differences therefore between the vegetation of the uncultivated parts of the Coal-measures and the other siliceous strata are due almost entirely to altitude, and not to any differences either in the nature of the climate or of the soil.

[1] The first number in brackets after an author's name refers to the year of publication in which the book or paper, which is being referred to, was published. The number after the colon refers to the page where the particular matter which is referred to occurs in the original work. The pages of the quoted works refer, as far as is possible, to the pages of the original memoirs, and not necessarily to the pages of the separately issued copies, as these unfortunately are often paged differently from the original. The titles, dates, and places of publication of the works quoted will be found in an appendix (pp. 222–229).

The hill slopes of the Coal-measures are rarely very steep, and are characterized by a number of typical oak (*Quercus sessiliflora*) woods which are at least semi-primitive in character. These woods occasionally spread out for a little distance on to the low, flat or gently sloping plateaux. The valley bottoms are almost filled up with overgrown, manufacturing villages, many of which have a population of twenty thousand people. The smoke from the villages and hamlets, for even every hamlet has its factory, frequently renders the sky dull and the atmosphere thick and heavy. Sandstone walls as a rule, hedgerows occasionally, separate the cultivated fields; and the stone walls and the tree trunks are permanently blackened with soot and smoke which have effectually destroyed almost all traces of mural plants, especially Cryptogams. Only in the heart of the woods, some of which retain their original sylvan character, may one, in this Coal-measure country, forget the propinquity of coal-mines and mills.

The higher hills of the central *massif* consist of sandstones and shales belonging to the Millstone Grit and to the Yoredale or Pendleside (Hind, 1897, etc.; Hind and Howe, 1901) series of rocks. Here there are fewer factories than on the Coal-measures, and no coal-pits. The higher hill summits are unpopulated, and covered with peat moors. Here and there, the moorland plateaux terminate abruptly in precipitous escarpments, locally known as "edges," formed of massive sandstone rocks. The larger and broader valleys are known as "dales," the smaller and narrower ones as "cloughs" or "deans," or, further north, as "ghylls." The upland valleys shelter woods of oak (*Quercus sessiliflora*), and rarely of birch (*Betula pubescens*); but more frequently the slopes of the steep valleys are tenanted by scrub or grassland. The bracken is a characteristic plant of the drier slopes. The upper portions of the cloughs contain numerous reservoirs (see figure 36) which are fed by the streams issuing from the peat moors of the plateaux. The lower plateaux and valleys are cultivated, chiefly as permanent pasture: arable land is decidedly scarce: wheat, in particular, is very rarely grown; and even fields of oats are uncommon. The fields are usually separated by sandstone walls; though, as in the Coal-measure country, hedgerows occur where the shales are of great extent.

In the south of the district, a third type of scenery is occasioned by the rocks of the Carboniferous or Mountain Limestone. The limestone plateaux are not so high as those of the sandstones; and they are frequently cultivated up to their summits. Limestone escarpments are frequent, and are more or less covered with plants, many of which belong to quite different species from those which characterize the sandstone escarpments. The valleys are all spoken of as "dales"; and these are much richer in species than the "cloughs" of the sandstones and shales. The valley slopes are steep, and are clothed by ash (*Fraxinus excelsior*) woods, or scrub, or calcareous grassland. The limestone country is too remote from the factories to be affected seriously by smoke. Arable land, on which oats are commonly grown, is not rare; but wheat is practically never grown on the Mountain Limestone. The fields are separated by white, limestone walls which give to the country side a very characteristic appearance.

Generally speaking, the cloughs in the shaly areas are grassy: those of the sandstone areas are bolder, more rocky, and more heathery. The prevailing hues of the cloughs are warm browns and purples, those of the limestone dales cold greys and greens, for in the latter localities, bracken, heather, and bilberry are almost entirely absent.

Rocks and Soils

The geological features of the district have been elucidated by Green (1869 and 1887), Dale (1900), and others. Still, the features of a district which are of chief interest to the geologist are not necessarily those which are responsible for the differences of the vegetation. From the latter point of view, it is the soil that is important (cf. figures 1 and 2); and this is not always directly related to the solid strata that are indicated on an ordinary geological map. In the present district, although it is largely unglaciated, there are several important soil features which cannot be inferred from any of the existing geological maps. Unfortunately, only geological maps of the old series are issued for this district; and no soil maps and no drift maps of the Peak District have been published by the Ordnance Department. In fact, the survey of the drift of this district does not appear to have

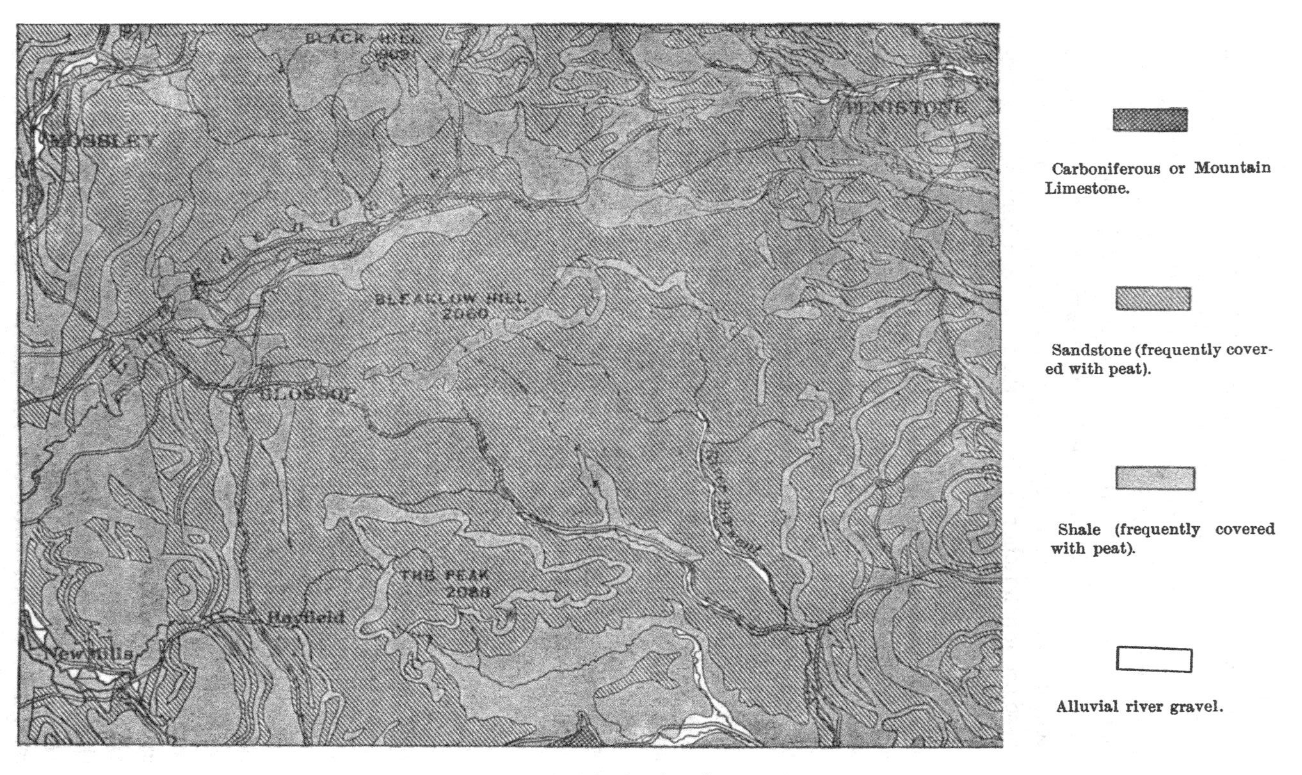

Figure 1. Subsoil map of the Peak District (northern area).

been yet commenced by the Geological Survey, though several papers on the subject have been published by various geologists.

Considering first those soils which are directly a result of the underlying strata, it is, so far as this district is concerned, necessary and sufficient to distinguish two main classes, calcareous and non-calcareous. The latter overlie the rocks of the Pendleside series, the Millstone Grits, and the Coal-measures. The calcareous soils overlie most of the rocks of the Carboniferous Limestone.

The non-calcareous soils of the Pendleside beds, the Millstone Grits, and the Coal-measures are here taken together, for the soils to which they directly give rise are essentially alike in their chemical and physical characters, and produce identical plant associations. The rocks consist largely of coarse grains of sand, of pebbles, of quartz, of pieces of decomposed felspar, and of flakes of mica.

Local floristic differences may perhaps be related to the different strata; but, in any case these differences are very slight. Linton (1903: 15) correctly states that the Coal-measures "can scarcely be said to possess a distinctive flora"; but the plants given by Linton (*op. cit.*) as characteristic of the grit are, in nearly all cases, plants confined to peat; and he gives no list of species characteristic of the Pendleside (or Yoredale) rocks, erroneously including these with the Carboniferous Limestone. The beds of all three series of rocks consist of alternating beds of sandstones and shales. In no other part of the British Isles are these strata so characteristically developed or so widespread as in the region of the Pennines. Over the shales, the surface soils weather ultimately into a kind of false clay, dark yellow in colour, and very slippery when wet. The soils produced by the weathering of the sandstones consist, when newly formed, of yellow sand; but this quickly becomes mixed with humus, when its colour is much darker. Pure sand is of very limited occurrence in the Peak District, and is almost limited to the vicinity of quarries, where a few arenicolous, as opposed to silicolous, species sometimes occur, such as *Spergularia rubra.*

Generally, the soils over the sandstones and shales are poor in soluble mineral salts, especially calcium carbonate. Woodhead (1906: 376) states that soils of this type in the Huddersfield district only contain from 0·02 to 0·04 per cent. of lime. The soil is usually rich in humus, and therefore retentive of water. Over such soils, if left uncultivated and undisturbed, peat inevitably develops in course of time.

The sandstones and shales are usually regarded as having been originally formed from the waste and denuded material of a great tract of granite. The resulting soils are of a siliceous nature, very deficient in soluble mineral salts, whilst in texture they are intermediate between loam and clay. The soils are shallow, as in the case of practically all siliceous soils derived from the Palaeozoic rocks; and the most typical vegetation consists of grassland dominated by the mat-grass (*Nardus stricta*) and the silver hair-grass (*Deschampsia flexuosa*).

There is a popular but quite erroneous impression that the soils over the rocks of the Pendleside (or Yoredale) series of the southern Pennines are calcareous; and, in Linton's *Flora of Derbyshire* (1903), the plant records are partly arranged on this assumption. The error may perhaps be accounted for by the fact that the true Yoredale rocks of the northern Pennines are frequently calcareous, and by the additional fact that, on the existing Ordnance maps of the Geological Survey on the scale of a quarter of an inch to the mile (1 : 253,440), the rocks of the Pendleside series and those of the Carboniferous Limestone are indicated by the same colour. It is true that the Pendleside rocks of the southern Pennines occasionally show thin bands of calcareous nodules; but these bring about little or no change in the vegetation.

The soil over the Carboniferous or Mountain Limestone is, in general, strongly calcareous, as this rock is composed very largely of molluscan shells, encrinites, and corals; but it agrees with that over the sandstones and shales in often being highly ferruginous, and in giving, from place to place, a great range of variation in water content. The highest percentages of calcium carbonate occur on the steep hill slopes; and this is no doubt due to the continuous exposures of new surfaces by denudation. The lowest percentages occur on the flatter plateaux; and this is doubtless caused by the leaching of the upper layers of the soil, the lime being carried away in solution to the subterranean or telluric waters, which find a ready means of escape to lower levels by means of the open joints of the limestone.

Many of the plateaux marked on the geological maps as consisting of limestone are capped by a layer of non-calcareous chert (cf. Sibley, 1908); and such plateaux yield soils which are essentially identical with those over the sandstones and shales. Sometimes the soil contains a mixture of stones of the limestone and of the non-calcareous chert; and then lime-loving plants occur. This agrees with the observations of Stebler (1906) in Switzerland.

Contemporaneous igneous rocks (cf. Arnold-Bemrose, 1907) occur in the limestone area. Although of comparatively limited extent, they are interesting locally. For example, a small patch of bilberry (*Vaccinium Myrtillus*) and of other lime-avoiding plants occurs on an outcrop of volcanic "toadstone" or basalt near Miller's Dale railway station, and is surrounded by lime-loving plants, *e.g.*, the salad burnet (*Poterium Sanguisorba*) growing on the limestone soil.

Of soils composed of recent deposits, there are the glacial sands, the river alluvia, and the upland peat.

The glacial drift of this district is confined to its western boundary. Boulder clay scarcely occurs; but non-calcareous, fluvio-glacial sands form rather extensive deposits, chiefly near the confluence of the rivers Etherow and Goyt. These deep and non-calcareous sands bring about a noteworthy change in the vegetation, as, in this district, woods of the pedunculate oak (*Quercus Robur* = *Q. pedunculata*) occur on this soil alone. The sands do not appear to occur much higher than about 600 feet (183 m.). To the west of the Peak District, on the plain of Lancashire and Cheshire, extensive glacial deposits are found, which consist largely of boulder clay, gravel, and sand. These deposits occur intermittently up to the crests of the hills which face the western plain, and also up the river valleys. For example, glacial boulders are to be found on the summit of Spond's Hill, at 1350 feet (411 m.); and they also occur in the valley of the Goyt, on the watershed, and in the valleys of the Wye and the Dove (cf. Dale, 1900, etc.). The boulders, however, are local in their occurrence, and bring about no appreciable change in the vegetation. Except on its western fringe, as on Tintwistle Moor, near Glossop, the general moorland plateau of the Pennines south of the Aire and Calder watershed is not glaciated. No perched blocks occur, no striae, and no *roches moutonnées*. It is not likely that traces of glaciation once existed here and have been obliterated, as the moorland plateau consists of uninhabited and unenclosed land where there is no necessity to remove boulders. Moreover, on hills immediately to the west, *e.g.*, on the Macclesfield moors, and on the moors some miles to the north, *e.g.*, on the Ilkley moors, glacial drift, boulders, and striae are found; and it is inconceivable that all traces of glacial action should have been entirely obliterated from the moors of the central and eastern Peak District, and not from the similar and neighbouring moors of Macclesfield and Ilkley. It is highly probable, then, that the Peak of Derbyshire and the high lands to the north, east, and south of the Peak, stood up, even during the time of maximum glaciation, as a *nunatak*, and that the ice-sheet fringed the hills of the west of the district. The fluvio-glacial sands are probably attributable to material washed out at the edge of the waning ice-sheet. Barrow (1903: 42) maintains that the glaciation of the neighbouring district of Cheadle, Staffordshire, ceased much earlier than in Northumberland and Scotland.

River alluvium, consisting generally of gravels, occurs at the bottom of most of the larger valleys. The gravels are non-calcareous in the valleys of the sandstones and shales, as, for example, between Hope and Grindleford, and calcareous in the limestone area, as, for example, in lower Monsal Dale. They bring about no important changes in the vegetation. In lower Monsal Dale, a calcareous alluvial flat is uncultivated, and the plants there are such as occur on the other calcareous soils; and near Grindleford, where a non-calcareous alluvial plain is also uncultivated, the plants are such as occur on the other non-calcareous soils. At the present time, the river gravels are mostly under cultivation, chiefly as permanent pasture; but a moderate quantity of wheat is grown on the gravelly alluvium near the confluence of the two streams, the Noe Water and the Derwent. In early times, it is not improbable that these alluvial tracts were characterized by woods of the "alder and willow series" (cf. Moss, Rankin, and Tansley, 1910: 122, *et seq.*).

Peat occurs on the summits of the higher non-calcareous hills, including the plateaux of chert in the limestone area, and is fully dealt with in Chapter VII. It is remarkable that very extensive deposits of peat in this country, both lowland

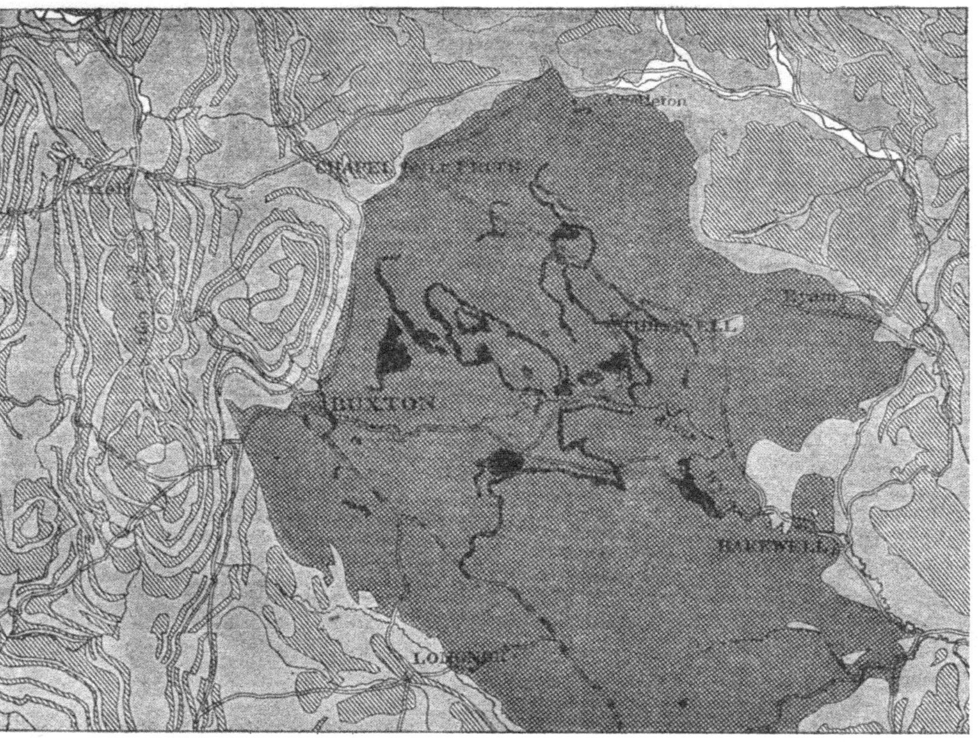

Figure 2. Subsoil map of the Peak District (southern area).

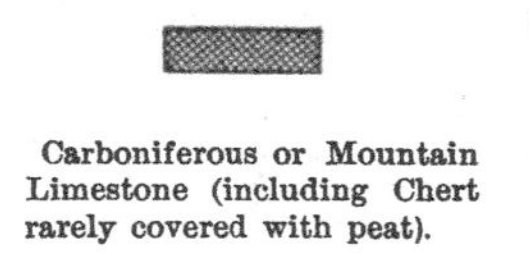
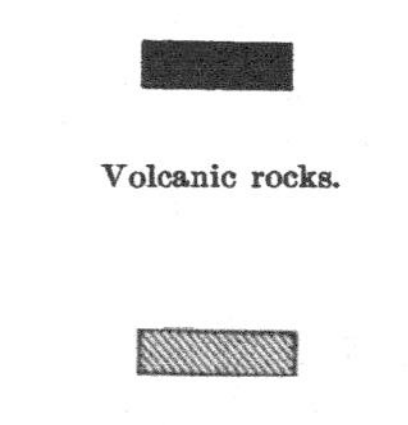
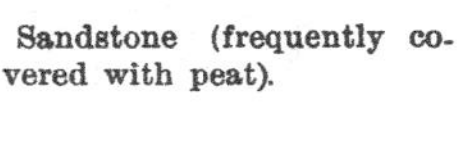

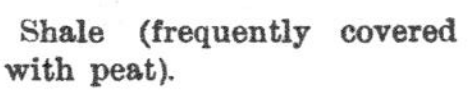

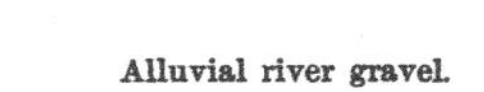

Carboniferous or Mountain Limestone (including Chert rarely covered with peat).

Volcanic rocks.

Sandstone (frequently covered with peat).

Shale (frequently covered with peat).

Alluvial river gravel.

peat and hill peat, should be ignored on most of the maps of the Geological Survey, even on their published drift maps. However, this fact makes such vegetation maps as accompany this memoir all the more valuable; as, on vegetation maps, the plant associations characteristic of peaty soils are indicated; and thus the occurrence and distribution of peat may be inferred.

The following table summarizes the chief strata and soils of the district, and states the general character of their accompanying vegetation. Farmland and plantations occur, to a greater or lesser extent, on all the rocks and soils, except on wet peat, and are omitted from the table. Marsh vegetation and stream vegetation also occur on all the soils; and these also are omitted.

Geological Strata	Soils	Vegetation
I. Carboniferous or Mountain Limestone rocks	1. On steep slopes, especially those below 1000 feet (305 m.):—shallow, brown, ferruginous, calcareous marls or gravels	1. Ash woods, scrub, calcareous grassland
	2. On more or less level ground, especially at elevations greater than 1000 feet:—shallow, dark brown, highly ferruginous, and more or less leached marls	2. Scrub, calcareous grassland, calcareous heath
b. Volcanic rocks	Non-calcareous soils varying from soft gravels to a kind of false clay, light brown or black in colour	Siliceous grassland
c. Chert	(*c* i) Non-calcareous soils varying from hard gravels to a kind of false clay, reddish brown in colour, or black when mixed with much acidic humus	(*c* i) Siliceous grassland, siliceous grassland mixed with heather
	(*c* ii) Peat, rarely exceeding one foot (30·5 cm.) in depth	(*c* ii) Heather moor, heather moor with much cotton-grass
d. Refuse heaps ("rakes") of old lead mines and of recent spar and gravel pits	Loose calcareous and cherty gravels containing salts and oxides of lead	Open plant associations usually characterized by the lead-wort (*Arenaria verna*) in abundance

Geological Strata	Soils	Vegetation
II. Pendleside [Yoredale] rocks:— A. Sandstones B. Shales III. Millstone Grit rocks:— A. Sandstones B. Shales IV. Coal-measure rocks:— A. Sandstones B. Shales	Non-calcareous, ferruginous sands, gravels, and shales, which at the surface degenerate into a kind of false clay, or which may be mixed with much acidic humus: the colour varies from yellowish brown to black according to the amount of humus present	Birch (*Betula pubescens*) woods and scrub, oak (*Quercus sessiliflora*) woods and scrub, siliceous grassland, siliceous grassland with much heather
V. Recent soils:—		
a. Glacial sands	(*a*) Yellowish non-calcareous sands, often mixed with acidic humus	(*a*) Oak (*Quercus Robur*) woods, scrub, siliceous grassland
b. River gravels	(*b* i) Yellowish, non-calcareous, sandy gravels	(*b* i) Stream-side scrub and siliceous grassland
	(*b* ii) Calcareous gravels and tufa	(*b* ii) Stream-side scrub and calcareous grassland
c. Upland peat	(*c*) Brown or black peat, of a depth of 15 feet (457 cm.) or more, and usually very wet	(*c*) Heather moors, cotton-grass moors, bilberry moors, retrogressive moors

Soils and their Characteristic Plants

It will be seen that the soils of the southern Pennines fall into three main types, namely, calcareous soils, siliceous soils, and peaty soils. The ground waters of the calcareous soils are alkaline in reaction, those of the siliceous soils neutral or acid, and those of the peaty soils of this district always acid. The siliceous and peaty soils of the district are more closely related to each other than either is to the calcareous soils, not only by the acidity but also by their low, soluble mineral-content and by their floristic composition. There is also every stage of transition from acidic siliceous soils to the acidic peaty soils; and in the following table of plants characteristic of (though not in all cases absolutely confined to) each kind of soil, several species are necessarily given as characteristic both of siliceous and of peaty soils.

Calcareous soils	Siliceous soils	Acidic peaty soils
Asplenium viride	Equisetum sylvaticum	"Lycopodium alpinum"[1]
A. Trichomones	"Osmunda regalis"	"L. inundatum"
A. Ruta-muraria	Pteris aquilina	Blechnum spicant
A. Adiantum-nigrum	Cryptogramma crispa	
Ceterach officinarum	Athyrium Filix-foemina	
Phyllitis Scolopendrium	Nephrodium montanum	
Cystopteris fragilis		
Polypodium vulgare		
Phegopteris Robertiana		
Taxus baccata		
"Juniperus communis"		
Silene nutans	Salix repens	Salix repens
Arenaria verna	S. aurita	Ranunculus Flammula
Helleborus viridis	Betula pubescens	*var.* tenuifolius
H. foetidus	Quercus sessiliflora	Drosera rotundifolia
Ranunculus circinatus	Montia fontana	Potentilla erecta
R. fluitans	Spergula arvensis	Rubus Chamaemorus
R. trichophyllus	Spergularia rubra	Genista anglica
R. Drouettii	Stellaria uliginosa	Ulex Gallii
Thalictrum minus	Ranunculus Flammula	Empetrum nigrum
T. flavum	R. Lenormandi	Viola palustris
*Cheiranthus Cheiri	Corydalis claviculata	
Arabis hirsuta	Rubus spp.	
*A. albida	R. Idaeus	
"A. perfoliata"	Potentilla erecta	
Cardamine impatiens	P. procumbens	
Draba muralis	Cytisus scoparius	
D. incana	Genista anglica	
Erophila verna	Ulex Gallii	
E. praecox	Polygala serpyllacea	
"E. inflata"	Empetrum nigrum	
Cochlearia alpina	Ilex Aquifolium	
Thlaspi alpestre	Hypericum humifusum	
var. sylvestre	Viola palustris	
var. virens		
*Iberis amara		
Hutchinsea petraea		
Sedum acre		
"Saxifraga sphonhemica"		
S. Telephium		
S. hypnoides		
S. tridactylites		
Parnassia palustris		
Ribes alpinum		
Spiraea Filipendula		
"Potentilla verna"		
Poterium Sanguisorba		

[1] Throughout this work, the names of species included within inverted commas are taken from Linton's *Flora of Derbyshire* (1903): names of species which are not indigenous are preceded by an asterisk.

Calcareous soils	Siliceous soils	Acidic peaty soils
Rosa spinosissima Pyrus Aria Anthyllis Vulneraria Hippocrepis comosa Ononis spinosa Geranium lucidum G. sanguineum Euonymus europaeus Rhamnus catharticus Hypericum hirsutum H. montanum Helianthemum Chamaecistus Viola hirta V. sylvestris V. Riviniana *var.* villosa Cornus sanguinea		
Ligustrum vulgare Polemonium coeruleum Myosotis collina Lithospermum officinale Origanum vulgare Thymus Serpyllum Satureia Acinos S. Calamintha Atropa Belladonna Verbascum Thapsus Plantago media "Rubia peregrina" Galium sylvestre *var.* nitidulum Asperula cynanchica Valerianella olitoria V. carinata Dipsacus pilosus Scabiosa Columbaria Campanula Trachelium C. glomerata Inula vulgaris Arctium nemorosum Carduus nutans Cirsium eriophorum Centaurea Scabiosa Picris hieracioides Hieracium spp.	Vaccinium Myrtillus Erica cinerea Calluna vulgaris Myosotis repens Scutellaria minor Digitalis purpurea Galium saxatile G. Witheringii Valeriana sambucifolia Jasione montana Wahlenbergia hederacea Gnaphalium uliginosum G. sylvaticum Senecio sylvaticus	Andromeda Polifolia Arctostaphylos Uva-ursi Vaccinium Myrtillus V. Vitis-idaea V. Oxycoccus Erica cinerea E. Tetralix Calluna vulgaris
Avena pubescens A. pratensis Koeleria cristata (agg.) Melica nutans	Agrostis canina Holcus mollis Deschampsia flexuosa Molinia caerulea	Potamogeton polygonifolius Agrostis canina Molinia caerulea

Calcareous soils	Siliceous soils	Acidic peaty soils
Bromus erectus	Nardus stricta	Nardus stricta
Brachypodium pinnatum	Carex helodes	Eriophorum vaginatum
Scirpus compressus	C. binervis	E. angustifolium
Carex disticha	C. echinata	Scirpus caespitosus
C. ornithopoda	C. panicea	Carex canescens
Allium vineale	C. Goodenowii	C. echinata
A. oleraceum	C. flava (agg.)	C. panicea
Polygonatum multiflorum	Luzula maxima	C. Goodenowii
P. officinale	L. multiflora *forma* congesta	C. flava (agg.) *var.* minor
Convallaria majalis	Juncus effusus	Juncus squarrosus
Paris quadrifolia	J. supinus	Narthecium ossifragum
"Cypripedium Calceolus" (extinct)	Orchis ericetorum	"Listera cordata"
Ophrys apifera		
O. muscifera		
Orchis ustulata		
O. pyramidalis		
Helleborine atrorubens		

Flora and Vegetation

The Peak District has been well studied from the floristic standpoint, for each county represented, with the exception of the small portion which occurs in south Lancashire, has its published local flora. Those of west Yorkshire (Lees, 1888), Cheshire (de Tabley, 1899), and Derbyshire (Linton, 1903) may be taken as representative, in various ways, of the best of the British county and local floras; and those by Painter (1889) and Bagnall (1901) provide useful lists of species. The flora by Crump and Crossland (1904), although it deals with an area a little to the north of this district, may be taken as illustrating the flora of the non-calcareous soils of the southern Pennines generally; and it shares, along with Wheldon and Wilson's flora (1907), the honour of being one of the very few British floras which deal at some length with the occurrence and distribution of plant associations.

Whilst, however, the flora of the Peak District has been dealt with by several authors, its vegetation has not been described. The distinction between flora and vegetation was emphasized by many of the older plant geographers, notably by Humboldt, A. P. de Candolle, Grisebach, and Thurmann. Thurmann stated (1849: 22) that "la Flore s'entend surtout du nombre des formes végétales distinctes qu'on y observe, la Végétation de leurs proportions et de leur association." From a flora, a knowledge is gained of the occurrence and distribution of the species of a district, of their presence or absence in contiguous districts, of the stations of these species, of the general nature of their habitats, of the altitudes to which they ascend, of their comparative abundance or rarity, of their times of flowering, and of their rank (*i.e.*, whether they are indigenous or not). So much may be expected of any flora which has pretensions to be a scientific work. In some floras, hints are given as to why certain species are confined to certain kinds of habitats: mention is made of those which are dominant over particular tracts of country: the floristic (not merely the topographical) subdivisions of the district covered by the flora are outlined; and some idea is given of the original migrations of the species into the district in question. Some modern floras rightly furnish details with regard to the very closely allied or "elementary species" which occur in the district, and state how these may be distinguished, whether or not their characters appear to be constant, and whether or not the plants in question are confined to special habitats.

The flora is composed of the individual species: the vegetation comprises the groupings of those species into *ensembles* termed vegetation units or plant communities.

A botanist who frequently traverses any stretch of uncultivated land, such as the elevated lands of the Peak District, must recognize sooner or later that the plants have become arranged in definite vegetation groups or plant communities. For instance, in the present district, the gentle slopes of the edges of the peat moors are almost entirely monopolized by heather (*Calluna vulgaris*), the higher peat moors by cotton-grass (*Eriophorum vaginatum*), and the highest and most exposed ridges by bilberry (*Vaccinium Myrtillus*). Here then he may distinguish three plant associations which he may term respectively heather moor, cotton-grass moor, and bilberry moor. These associations he finds to be constant both as regards their floristic composition and their general life conditions. Similarly, he may easily recognize other plant associations on the hill slopes of the district.

VEGETATION MAPS

In this way the observer determines that certain plant associations are typical of certain limited areas; and it becomes possible to construct vegetation maps on which the distribution of these associations may be shown. The number of the plant associations which may be indicated on a map depends very largely upon its scale. The bigger the scale of the map the more the plant associations which may be shown upon it; and considerable judgment is required in deciding which associations shall be shown on a map of any given scale. In deciding this difficult but very important question, several general considerations must be borne in mind.

The object of a vegetation map of a district under investigation is to give the best possible cartographical representation of the plant communities which the scale of the map will allow. On the one hand, the fullest advantage must be taken of the size of the scale employed; and, on the other hand, the map must not be so crowded with details that it loses in definiteness and clearness (cf. Flahault and Schröter, 1910: 11). The experience of phytogeographers in this country is that the scale of one inch to the mile (1: 63,360) is a suitable one for maps intended to show the distribution of the more important plant associations of the British Isles, and that maps of a smaller scale are not desirable except for special purposes.

It is obvious that every plant community cannot be indicated on a map of this scale (1 : 63,360); and hence the plant geographer has frequently to subordinate minor units of vegetation to units of wider significance; and, in such cases, the plant geographer has to determine the larger vegetation units to which the minor units must be subordinated: otherwise, the colours on a vegetation map will be mere empiricisms and without any philosophical basis. It is obvious, therefore, that no one can successfully construct a really scientific vegetation map unless he has specially considered the interrelationships of the fundamental units of vegetation.

Plant Communities

A *plant formation* is the whole of the vegetation which occurs on a definite and essentially uniform habitat. A *plant association* is of lower rank than a formation, and is characterized by minor differences within the generally uniform habitat. A *plant society* is of lower rank than an association, and is marked by still less fundamental differences of the habitat. These grades of difference of the habitat are marked by corresponding differences in floristic composition. The three vegetation units may be compared respectively with the taxonomic units of the genus, the species, and the variety; and differences of opinion arise, both among systematic and geographical botanists, with regard to the precise limits of the respective units. *Sub-formations* and *sub-associations* may be recognized; and these would be comparable to subgenera and subspecies. *Plant community* is a convenient and general term used for a vegetation unit of any rank.

Examples of the above types of plant communities are well illustrated in the present district. The acidic peat of the non-calcareous plateaux of the district furnishes a definite and generally uniform habitat; and the whole of the vegetation of this habitat is the plant formation of the acidic peat moors. The plant associations of *Calluna vulgaris* (*i.e.*, heather moor), of *Eriophorum vaginatum* (*i.e.*, cotton-grass moor), and of *Vaccinium Myrtillus* (*i.e.*, bilberry moor) are subdivisions of this plant formation, and are characterized by minor differences within the generally uniform habitat and by corresponding differences in floristic composition. Still less important differences in the habitat may cause certain members of the associations to become locally dominant; and thus, in the heather moor, for example, plant societies, or mere local aggregations of species, occur of *Erica cinerea*, of *Juncus squarrosus*, of *Carex Goodenowii*, and of other species.

For the purpose of vegetation maps on a scale of one inch to the mile (1 : 63,360), the most important vegetation unit is the association, as it is, in general, possible to indicate the more important associations on a map of this scale. To indicate the smaller associations and the plant societies, maps on a scale of six inches to the mile (1 : 10,560) are desirable; and, on maps whose scale is a quarter of an inch to the mile (1 : 253,440), plant formations could, in general, only be shown. It is extremely doubtful if vegetation maps of limited areas on a scale much smaller than this can be constructed on a strictly scientific basis until some method of classifying formations into larger but natural vegetation units has been devised.

A plant association is a *closed association* when the ground is fully occupied by plants, and when it is dominated either by a single species, as in the case of a heather moor, or by two or more species all belonging to the same plant form, as in the case of some reed swamps. A plant association is an *open association* when the ground is only partially covered with vegetation, as in the case of denuding peat moors. Other associations are intermediate in character between open associations and closed associations. In an *intermediate association*, the ground may be more or less fully covered with plants; but there is no single dominant plant or plant form: there are, in fact, several plants which compete with each other for dominance, as on an East Anglian fen, where *Cladium Mariscus, Phragmites communis, Molinia caerulea, Calamagrostis canescens, Juncus subnodulosus,* and other plants compete in this way. Differences in a single association caused by the varying abundance of the constituent species may be spoken of as the *facies* of an association. When, in a plant association, the more abundant species become very conspicuous at different times of the year, *seasonal aspects* of associations are produced (cf. Clements, 1905: 296 and 315).

If the succession of associations within a single formation is studied, it is found that the initial stages are marked by open and unstable associations, that these are followed by intermediate associations, and these again by stable associations (cf. Clements, 1904: 135; Moss, 1907 *a*: 12). The stable associations, however, may degenerate, and give rise to other intermediate associations. In the present account of the vegetation of the Peak District, the associations will be considered from this point of view; and accordingly the following terminology will be used. Open and intermediate associations leading up to a stable association are termed *progressive associations*: intermediate and open associations resulting from the decay of a stable association are termed *retrogressive associations*: progressive and retrogressive associations together are termed *subordinate associations*; and the stable associations are termed *chief associations*. The latter are virtually the "climatic formations" of Cowles (1911: 161), but not those of Schimper (1903: 161).

"Every formation has at least one chief association: it may have more; and they may be regarded (cf. Drude, 1896: 286) as equivalent to one another in their vegetational rank. They are more distinct and more fixed than progressive or retrogressive associations. They are usually, but not invariably[1], closed associations. They always represent the highest limit that can be attained in the particular formation in which they occur, a limit determined by the general life conditions of the formation. In desert and sub-niveal regions, the chief associations are open[1]; and, in such cases, it is legitimate to speak of open formations. Open progressive and retrogressive associations, however, frequently occur in formations whose chief associations are closed" (Moss, 1910 *b*: 38).

Every part of a plant formation necessarily belongs either to one of its subordinate associations or to one of its chief associations.

A plant association, whether open or closed, which is characterized by a single dominant species, is spoken of as a *pure association*, one characterized by several species competing for dominance as a *mixed association*.

A plant formation has a life-history. It is born: it enters on a period of infancy and adolescence, that is, of progressive associations: it reaches a period of maturity, that is, of chief associations: it passes through a period of senility or decay, that is, of retrogressive associations; but throughout these stages, it is the same organism characterized by a definite habitat which is related to a correspondingly definite flora.

Much discussion has taken place as to whether or not a particular plant formation may be world-wide in its distribution. From the point of view here taken, climatic factors and geographical position are regarded as part of the habitat; and it follows that any particular plant formation is confined to a single climatic or geographical region.

[1] Hence the statement that "for each habitat there is a closed, ultimate, or chief association" (*Bot. Centralblatt*, 1911: 100) is erroneous.

Each of the succeeding chapters deals with a group of associations, not necessarily with a formation. The associations are analysed; and, as far as is possible, each association is then referred to the formation to which it belongs.

Vegetation Maps and Floristic Maps

Vegetation maps indicate the occurrence and distribution of plant communities. Floristic maps may be of two kinds: they may indicate the occurrence and distribution of single species or of groups of geographically related species. The former maps are of the type which H. C. Watson (1832, etc.) began to construct of the species indigenous to the British Isles. They are very useful maps in their way, as may be seen by the use to which they are put in Praeger's recent *Flora of the West of Ireland* (1909: figs. 4, 5, 14, etc.). Such maps, however, do not lend themselves to any generalized cartographical scheme, because almost every species requires a separate map to show its distribution. They bear the same relation to vegetation maps that a series of cartographical representations of erratic boulders would bear to a modern geological drift map. Floristic maps illustrating the distribution of geographically related species can hardly be said to exist as regards the British Isles; but Flahault (1901) has constructed such a map of France, and more recently Massart (1910) has published maps of Belgium on somewhat similar lines.

The so-called "botanical maps" illustrating numerous British county and local floras are neither floristic maps nor vegetation maps. The typical maps accompanying these floras simply show divisions of the county into "drainage districts" or other topographically convenient districts; and no attempt is made on such maps to show the occurrence and distribution either of plant communities or of floristic groups of species.

The Value of Vegetation Maps

Vegetation maps have the same value to botanists that geological maps have to geologists. Just as geologists may, by consulting geological maps, know where certain geological phenomena may best be studied, so botanists may, by consulting vegetation maps, know where certain ecological phenomena may best be investigated. Geological maps show the nature and distribution of the chief rock strata; and vegetation maps indicate the nature and distribution of the principal plant communities. Geological memoirs describe the development and structure of the strata shown on the geological maps, and give lists of the fossils found in the deposits; and similarly vegetation memoirs give accounts of the development and structure of the plant communities, and furnish lists of the species which constitute the various units of vegetation.

The nature of the surface soil of a district may often be inferred from vegetation maps (cf. p. 12), even when the existing geological maps are not helpful in this regard, as is frequently the case.

Geographers too find the maps of service, as has recently been testified by Professor A. J. Herbertson, who states that "at last we have some modern botanical geography which is really valuable to the geographer" (Herbertson, 1911: 384).

The maps are also valuable to scientific agriculturists, who find on them the limits of profitable wheat cultivation mapped with very considerable accuracy (cf. pp. 204—5).

The forester may, from the nature of the natural and semi-natural woods shown on vegetation maps, obtain valuable data with regard to the prospects of success of planting certain indigenous or non-indigenous timber trees in any particular locality; and, to those interested in any future great scheme of afforestation, the vegetation maps which have been published will yield extremely valuable information (cf. p. 68).

Vegetation maps furnish the only reliable knowledge which is at present available with regard to the nature and possible utilization of the "waste lands" of the country. The Board of Agriculture has at its disposal an almost unlimited amount of information, much of which is published in their annual *Agricultural Returns*, with regard to the cultivated lands of the country; but, apart from unofficial vegetation maps and memoirs, there are practically no means of obtaining reliable knowledge of the nature and possible utilization of the uncultivated land of any portion of the British Isles.

Whilst, however, the geological survey of the country is carried on by public funds, the vegetation survey languishes under voluntary efforts. There are at present about twenty finished vegetation maps of different parts of Great Britain, such as the Isle of Wight, Hampshire, the Cleveland District of Yorkshire, and Lanarkshire; and these maps cannot at present be published owing to lack of funds. The time seems to be approaching, therefore, for placing a vegetation survey of this country on the same official basis as the existing geological survey.

RAINFALL[1]

There are not enough rainfall stations in the district, especially in the moorland region, with sufficiently long and continuous records to justify the construction of a map showing rainfall lines. Dr H. R. Mill, however, has kindly supplied the following statistics giving yearly and monthly averages for thirty years at ten stations distributed as regularly as possible over the area covered by the vegetation maps. The figures given on the following page represent the average rainfall to the nearest inch, during the thirty years 1870—1899. The particular rainfall stations have been selected by reason of the fact that they possess long and accurate records; and the yearly figures may be taken as satisfactory for the stations in question. A slightly greater rainfall may be safely assumed to occur on the higher grounds and on the leeward side of the highest hills. The figures showing the monthly averages may be taken as fairly satisfactory; but experience shows that for monthly rainfall figures to be fully satisfactory, fifty years or more are required, because a rainfall equal to the monthly average may occur in a single day, and because, on the other hand, a month may have no rainfall at all. The monthly rainfall of Burton-on-Trent (see the left-hand column) is added in order to furnish a comparison with a neighbouring town situated at a low altitude.

It will be seen from the figures in the table which follows that the first five months of the year are the driest, and that the driest of all is April, in spite of a popular opinion to the contrary. The soil, however, is often very wet during these months, owing to low evaporation. Of the remaining seven wet months, October is, in this district, by far the wettest.

[1] This section has been kindly revised by Dr H. R. Mill, the Director of the British Rainfall Organization, and Editor of *British Rainfall*.

Although rainfall stations are not numerous on the hills of north Derbyshire, Dr Mill (*vide* Linton, 1903 : 3) thinks that "average falls exceeding fifty inches (127 cm.) in the year occur at several points in the high ground; but it is unlikely that so great a rainfall as sixty inches (152 cm.) is reached."

Average Monthly Rainfall for the 30 *years,* 1870—1899

Burton-on-Trent	Months	Macclesfield	Chatsworth	Buxton	Whaley Bridge	Redmires	Arnfield	Woodhead	Ingbirchworth	Saddleworth	Meltham
in.		in.[1]	in.	in.	in.	in.	in.	in.	in.	in.	in.
2·11	January	2·81	2·92	4·47	3·33	3·37	3·20	4·07	3·18	3·50	3·63
1·64	February	2·29	2·46	3·38	2·85	2·85	2·62	3·50	2·76	2·57	2·76
1·59	March	2·69	2·35	3·80	3·04	3·00	2·92	3·71	2·80	3·07	2·91
1·85	April	2·07	2·14	2·75	2·33	2·82	2·25	3·09	2·50	2·37	2·46
2·21	May	2·78	2·42	3·33	2·78	2·92	2·71	3·35	2·61	2·54	2·54
2·67	June	3·38	2·84	3·62	3·22	3·18	3·17	3·66	3·02	2·77	3·19
2·73	July	3·83	2·83	4·14	3·77	3·44	3·57	4·20	3·28	3·59	3·41
2·90	August	3·86	3·19	4·56	3·96	3·54	4·01	4·44	3·23	3·95	3·79
2·62	September	3·80	2·92	4·44	3·77	3·55	3·86	4·65	3·48	4·03	3·81
3·07	October	4·26	4·52	5·55	4·50	4·90	4·64	5·85	4·66	4·64	4·79
2·31	November	3·39	3·37	5·00	3·82	3·81	3·58	5·06	3·77	3·71	3·78
2·31	December	3·51	3·26	5·13	3·99	3·81	3·68	4·60	3·60	3·74	3·93
28·01	Total	38·67	35·22	50·17	41·36	41·19	40·21	50·18	38·89	40·48	41·00

SMOKE

The district, lying as it does between the great coalfields and manufacturing districts of south Lancashire, and south-west Yorkshire, possesses an atmosphere which is frequently vitiated by smoke. The greasy soot settles on the leaves and stems of plants, and gives them a permanently dirty appearance. A clean expanse of white snow on the hills is often palpably blackened in two or three hours' time. Mr A. Wilson (1900) has stated that "the great smoke drift from south and east Lancashire could be seen crossing the Pennine Range of moorlands, and

[1] 10 inches = 25·4 centimetres.

then mingling with the West Riding smoke." Dr Cohen (1900, 1909) has investigated the smoke nuisance in great detail, and he emphasizes the view that the smoke lessens the amount of bright sunshine and lessens the temperature, even at some distance from the towns.

The reduction of light intensity by smoke does not, of itself, affect vegetation adversely, as, at ordinary temperatures and with the available amount of carbon dioxide, the natural illumination is greatly in excess of the amounts which plants utilize in their photo-synthetic activities; but the reduction of temperature which results from a smoke-laden atmosphere must adversely affect the carbon assimilation of plants and reduce their vitality. The clogging of stomata by grease and soot would also seem to be a factor of importance, as this must lessen the quantity of carbon dioxide which the plants can obtain; and it is known that plants are able to utilize more carbon dioxide than is ordinarily available. It is also known that in smoke, certain injurious acids, *e.g.*, sulphurous acid (H_2SO_3), are present, and that these have a poisonous effect on plants. Further, the soil is also injured by smoke on account of the sulphurous and other acids being deposited upon it. The injurious effects of fog on plants have been fully investigated and discussed by Professor F. W. Oliver (1893).

Whether or not the deleterious influence of smoke is a limiting factor as regards the plant associations of the district is doubtful, as it seems likely that all the associations may be affected to an equally adverse degree. On the other hand, Wilson (*l.c.*) shows that certain Cryptogams (*e.g.*, *Ulota* spp. and *Orthotrichum* spp. and lichens) are much rarer than formerly; and, in certain cases, he attributes this fact to smoke.

TEMPERATURE

Judging from the figures in the standard physical atlases (see Bartholomew and Herbertson, 1899), the mean annual temperature of the district is about 49° F. (9·44° C.). This may be compared with the mean annual temperature of Penzance (see Davy, 1909: xx) which is 52·68° F. (11·5° C.). January, with an average temperature of 36° F. (2·2° C.), is the coldest month of the year, and July, with an average temperature of

62° F. (16·6° C.), the warmest. According to Lees, the average daily range of temperature in the West Riding of Yorkshire (the south-west of which is included in the present district) is about twelve or thirteen degrees F. (7·2° C.) in the shade, and about thirty-two to thirty-four degrees F. (18·9° C.) under exposure to the direct rays of the sun (Lees, 1888: 6). These temperatures, it should be remembered, are corrected to sea level, and hence do not show the actual temperatures of the district.

Phenological observations show that the opening of spring flowers and the unfolding of the leaves of trees is from two to four weeks later on the Pennines than in the lowlands of Somerset, and from ten days to three weeks later than in Cambridgeshire.

The winter on the Pennine uplands is a long one, lasting, as a rule, from the beginning of November to the beginning of May, though snow does not, as a rule, lie very long on the ground, owing to intervals of mild weather. Although some moorland plants, such as the bilberry (*Vaccinium Myrtillus*), the crowberry (*Empetrum nigrum*), and the cotton-grasses (*Eriophorum vaginatum* and *E. angustifolium*) flower in late April and early May, the plants of the moorlands make very little new growth before June and July.

The Upper Atmosphere

I am indebted to Professor A. Schuster, F.R.S., for kindly supplying me with the reports (Schuster, 1908–9) for the years 1908 and 1909 of the investigation of the upper atmosphere conducted by the meteorological department of the Victoria University of Manchester. This investigation was begun on January 1st, 1908. The meteorological station is situated near Glossop, in Long. 1° 57 W. and Lat. 53° 24 N., at a height of 335 metres (1100 feet, approx.) above sea level. The details which are here utilized consist of readings of temperature, direction and velocity of the wind, and the humidity of the air, at altitudes respectively of 335 metres (1100 feet), 500 metres (1640 feet), and 750 metres (2460 feet). The readings of the higher altitudes were obtained by means of kites and balloons; and the figures utilized represent only a small proportion of the total number of observations made by Professor Schuster and his staff.

The readings at 335 metres may be taken as indicating, in a general way, the conditions which prevail at the present upper limit of woodland, in the scrub and grassland associations, and in the heather association. Those at 500 metres may similarly be regarded as illustrating the conditions which prevail in the cotton-grass association, and those at 750 metres the conditions a little above the bilberry and retrogressive moorland associations.

It is necessary to point out that most of the readings here reproduced were taken during the afternoons. Hence, as regards temperature, the figures roughly represent maxima. The figures relating to the direction and velocity of the wind may be taken as fairly satisfactory averages of the periods which they represent; and those relating to the humidity of the air also represent fairly satisfactory averages of readings taken during the daytime. It is obvious that observations are required for many more years before real averages may be deduced; but, so far as the figures go, some interesting results are indicated.

Temperatures

In accordance with expectation, there is a decrease in temperature as the higher altitudes are reached; but this decrease is greater than would have been predicted by Dalton's rough and ready rule. This rule states that the temperature falls 1° F. for every ascent of 300 feet (91·4 m.). However, from the figures shown on the next page, it may be calculated that, in 1908 and 1909, on the hills of the Peak District, the actual fall of temperature for every ascent of about 300 feet was very much greater than this. Of course, many more readings are necessary before any rule, which accurately represents the real phenomenon, can be stated.

December, January, February, and March are shown by the following tables to be cold months with combined averages of 2·6°, 1·8°, and 0·5° C. at the three altitudes respectively, during the two years. May, June, July, August, September, and October are warm months with similar averages of 12·9°, 10·9°, and 8·9° C.; and November and April are intermediate, with similar averages of 5·9°, 4·7°, and 2·7° C.

The coldest day of 1908 was December 29th, when temperatures of − 7·2°, − 7·8°, and − 9·4° C. were registered at the three altitudes respectively. In 1909, the coldest days were (1) February 12th, when −2·2°, −3·6°, and −5·6° C. respectively were registered, (2) March 4th, when −5·0° and −6·1° C. were

1908	At 335 metres		At 500 metres		At 750 metres	
	Days[1]	Degrees in C.	Days[1]	Degrees in C.	Days[1]	Degrees in C.
January	26	2·3	26	1·8	21	1·1
February	25	3·5	25	2·4	19	0·5
March	23	3·1	23	1·8	19	0·1
April	25	5·9	25	4·2	25	2·0
May	25	12·4	25	10·6	25	8·3
June	26	15·1	24	13·0	26	10·7
July	22	20·1	22	13·5	22	11·6
August	20	13·0	20	11·5	19	9·7
September	25	12·1	25	10·7	25	8·7
October	26	11·7	26	10·9	26	10·2
November	25	5·6	24	5·2	23	4·0
December	27	3·4	27	2·6	27	2·1
Average	25	9·0	24	7·4	23	5·8
1909						
January	25	2·2	23	1·4	23	−0·2
February	24	1·6	23	0·7	22	−0·1
March	27	2·2	26	1·2	19	−0·2
April	22	8·0	22	6·1	22	3·6
May	24	11·6	24	9·6	21	6·8
June	24	11·0	20	8·8	18	6·6
July	25	13·7	24	12·3	24	10·0
August	20	13·7	20	11·9	20	9·7
September	26	11·8	23	10·1	23	8·1
October	23	9·0	23	7·3	23	5·9
November	22	4·2	19	3·3	18	1·1
December	24	2·2	17	2·1	17	0·9
Average	24	7·6	22	6·2	21	4·4

[1] The numbers in these columns represent the number of days on which readings were taken. No readings were taken on Sundays; and occasionally readings could not be taken on other days, sometimes owing to accidents to the kites and balloons, and occasionally owing to requests from the landowner during the grouse-shooting season.

registered at the two lower altitudes[1] respectively, and (3) December 21st, when – 5·4° C. was registered at 335 metres[1].

On the days when observations were made, the temperature, during 1908, did not fall to zero (C.) from May to November inclusive, nor, during 1909, from April to October inclusive.

The following table shows the number of days during 1908 and 1909 when the temperature failed to rise above zero at an altitude of 350 metres:—

	1908	1909
January	11 days out of 26	7 days out of 25
February	2 days out of 25	12 days out of 24
March	1 day out of 23	7 days out of 27
April	2 days out of 25	0 days out of 22
November	0 days out of 25	2 days out of 22
December	6 days out of 27	6 days out of 24
Total	22 days out of 151	34 days out of 144

There is very little growth discernible in the vegetation of the Peak District at altitudes above about 1100 feet (335 m.) until the month of May: growth almost ceases in October; and from November to April inclusive, weather of a wintry type alternating with moderately mild intervals may be expected. These are much more severe conditions than exist in the lowlands, and are doubtless related to the meagre flora of the uplands, as the edaphic conditions of both the lowlands and the uplands of England in the latitude of the southern Pennines are practically identical.

Direction of the Wind

The figures in the tables on p. 31 are arranged to show, as far as is possible, the relative prevalence of warm and moist winds, *i.e.*, westerly and southerly winds (N.W. to S.S.E.) and of dry and cold winds, *i.e.*, northerly and easterly winds (S.E. to N.N.W.). It will be seen that the warm and moist winds prevail approximately on two days out of every three, and that (as might have been predicted) there is little change of direction at the three chosen altitudes. Direction of wind therefore is

[1] Readings at the upper altitudes on these days were incomplete.

	At 335 metres			At 500 metres			At 750 metres		
	Days	Direction of wind		Days	Direction of wind		Days	Direction of wind	
1908		N.W.[1] to S.S.E.	S.E. to N.N.W.		N.W. to S.S.E.	S.E. to N.N.W.		N.W. to S.S.E.	S.E. to N.N.W.
January	25	18	7	26	21	5	24	17	7
February	25	17	8	25	14	11	19	8	11
March	25	13	12	25	13	12	21	12	9
April	25	12	13	25	10	15	23	12	11
May	25	20	5	25	16	9	25	18	7
June	26	16	10	26	18	8	26	17	9
July	27	20	7	27	19	8	27	19	8
August	26	16	10	26	16	10	25	14	11
September	25	17	8	25	17	8	25	18	7
October	27	17	10	27	19	8	27	20	7
November	25	19	6	25	20	5	24	19	5
December	27	22	5	27	24	3	27	24	3
Total	308	207	101	309	207	102	293	198	95
1909									
January	26	22	4	26	22	4	24	20	4
February	24	13	11	24	13	11	24	13	11
March	27	14	13	27	14	13	25	12	13
April	26	21	5	26	20	6	26	20	6
May	25	19	6	25	20	5	25	19	6
June	26	10	16	26	9	17	25	8	17
July	27	24	3	27	22	5	27	22	5
August	26	22	4	26	21	5	24	20	4
September	26	14	12	26	13	13	26	12	14
October	24	21	3	24	20	4	24	21	3
November	22	14	8	21	11	10	21	9	12
December	23	14	9	22	15	7	22	15	7
Total	302	208	94	300	200	100	293	191	102

[1] To be read clockwise.

not to be regarded as a factor of importance in its relations to the differential distribution of the plant associations within a limited area, but rather as a factor which is fairly constant over extensive districts. As a limiting factor, therefore, it only becomes significant when different geographical or climatic districts are under comparison.

Velocity of the Wind

	At 335 metres		At 500 metres		At 750 metres	
1908	Days	Metres per second	Days	Metres per second	Days	Metres per second
January	26	4·9	26	10·3	21	14·4
February	25	6·6	23	13·2	19	15·1
March	23	4·4	24	8·0	19	8·5
April	25	5·4	23	9·4	21	10·7
May	22	5·3	20	10·0	20	11·6
June	26	4·6	23	10·3	23	9·6
July	23	4·3	23	8·4	22	10·8
August	26	5·4	25	8·6	24	10·1
September	24	5·3	21	10·3	21	13·6
October	27	5·3	24	9·8	24	11·5
November	23	5·6	20	9·5	19	12·4
December	27	4·2	24	8·2	24	10·6
Average	25	5·1	23	9·7	21	11·6
1909						
January	26	5·0	26	8·2	21	12·5
February	24	5·2	24	9·4	22	12·5
March	27	5·3	25	7·8	20	11·1
April	25	6·4	25	9·3	23	12·4
May	25	5·4	25	8·6	24	11·3
June	25	4·8	24	7·6	22	10·0
July	27	5·3	27	8·6	27	12·2
August	27	3·9	27	7·1	23	9·6
September	26	3·7	25	7·4	25	9·5
October	22	5·6	21	9·0	21	12·1
November	21	4·5	21	8·8	21	12·5
December	21	6·3	19	10·1	15	13·7
Average	25	5·1	24	8·5	22	11·6

Velocity of the Wind

	1908			1909		
	335 m.	500 m.	750 m.	335 m.	500 m.	750 m.
January	1*/26	5/26	11/21	0/26	1/26	9/21
February	1/25	6/23	†	0/24	1/24	7/22
March	0/23	1/24	†	1/27	1/25	4/20
April	0/25	3/23	5/21	1/25	2/25	4/23
May	0/22	3/20	4/20	0/25	0/25	3/24
June	0/26	1/23	1/23	1/25	1/24	4/22
July	0/23	0/23	1/22	0/27	0/27	5/27
August	0/26	1/25	2/24	0/27	0/27	3/23
September	0/24	2/21	6/21	0/26	0/25	3/25
October	0/27	0/24	2/24	2/22	3/21	8/21
November	1/23	0/20	1/19	0/21	2/21	8/21
December	0/27	0/24	1/24	0/21	4/19	6/15
Total	3/297	22/276	34/219	5/296	15/289	64/264

* The numerator represents the number of days on which the velocity of the wind reached 15 metres per second, and the denominator the number of days on which readings were taken.

† Observations defective.

The importance of wind velocity as regards vegetation is that transpiration increases with the velocity of the wind. It is not known whether the relation is in any degree precise, but the general fact would appear to be incontestable. It will be seen that there is a considerable increase in the velocity of the wind as the higher altitudes are reached; and this fact, especially when considered with the fact that the temperature decreases at the higher altitudes, helps to explain, in no inconsiderable degree, the paucity of the flora and the remarkable features of the vegetation of the higher altitudes. It must be remembered, also, that the peaty soil which characterizes nearly all the higher summits of the district is usually regarded as being physiologically dry; and this, if the fact be so, further increases the dangers of those plants which are exposed to excessive transpiration. The distribution of the days on which the velocity of the wind reached 15 metres per second is shown on p. 33.

Humidity of the Atmosphere

Whilst the records for the two years indicate a mean monthly humidity of the air of about 85 per cent., there are remarkable fluctuations in the daily records which are interesting when considered in relation to the conditions of plant life. The important point, generally speaking, with regard to atmospheric humidity, is that the lower the humidity the greater the amount of evaporation or transpiration. Schimper (1903: 4) states that transpiration "constantly increases in proportion to the dryness of the atmosphere." The daily records show that the atmospheric humidity fluctuates considerably at the different altitudes; and this variation is sometimes in the direction of greater humidity at the higher altitudes and at other times in the contrary direction. However, the lowest percentages occur at the higher altitudes on the whole; and, as plants, in order to survive, must be adapted to the extreme conditions of their habitats, it follows that plants at the higher altitudes are disadvantageously situated not only as regards a lower temperature and a greater wind velocity, but also, on the whole, as regards a lower humidity of the atmosphere. Whether or not this applies to altitudes much higher than 750 metres has yet to be determined.

1908	At 335 metres		At 500 metres		At 750 metres	
	Days	Per cent.	Days	Per cent.	Days	Per cent.
January	26	95	26	93	21	90
February	25	94	25	94	19	91
March	23	85	23	85	19	84
April	25	88	25	87	23	88
May	25	85	25	87	25	90
June	26	81	26	83	26	85
July	22	80	21	82	21	85
August	20	83	20	86	19	90
September	25	83	25	81	25	81
October	26	84	26	80	26	76
November	25	88	24	87	23	86
December	27	87	17	90	27	84
Average	25	86	24	86	23	86
1909						
January	26	85	24	84	22	84
February	24	83	24	82	23	81
March	27	90	27	90	25	90
April	23	71	22	71	22	73
May	25	66	25	65	22	63
June	23	79	19	78	19	85
July	26	84	25	86	25	87
August	24	88	23	88	23	90
September	26	89	23	90	23	86
October	25	86	24	86	24	88
November	21	91	18	92	18	92
December	23	96	17	96	17	94
Average	24	84	23	84	22	84

The lowest readings were registered on October 3rd, 1908, and on January 28th, 1909, when the humidity fell so low as 25 per cent. at 750 m. The month of May, 1909, is interesting on account of its low atmospheric humidity in this district. The monthly average at the three altitudes of this month was 65 per cent.; and, at 500 m., out of the 25 days on which readings were taken, the humidity fell below 60 per cent. on 12 days. The number of days on which the atmospheric humidity fell below 60 per cent., and the monthly distribution of these days, are indicated below:

	1908			1909		
	335 m.	500 m.	750 m.	335 m.	500 m.	750 m.
January	0*/24	0/26	2/21	3/26	3/24	3/22
February	0/25	1/25	1/19	2/24	3/24	5/23
March	0/23	1/23	1/19	0/27	0/27	0/25
April	1/25	2/25	2/23	3/23	4/22	5/22
May	1/25	1/25	0/25	9/25	12/25	10/22
June	2/26	1/26	1/26	0/23	0/19	1/19
July	2/22	2/21	1/21	0/26	0/25	1/25
August	1/20	0/20	0/19	0/24	0/23	0/23
September	0/25	1/25	3/25	0/26	0/23	0/23
October	0/26	1/26	2/26	0/25	0/24	0/24
November	0/25	0/24	2/23	0/21	0/18	0/18
December	0/27	2/17	3/27	0/23	0/17	0/17
Total	7/293	12/283	18/274	17/293	22/271	25/263

* The numerator represents the number of days on which the atmospheric humidity fell below 60 per cent., and the denominator the number of days on which observations were taken.

It will be seen that, whilst undertaken with a different object, the results of the observations on the upper atmosphere by Professor Schuster and his staff have several important bearings on vegetation, though the significance of these cannot be correctly appraised until the observations have been continued for a considerable number of years. Though admittedly incomplete, they still help to furnish clues as to the different conditions under which vegetation lives at the different altitudes.

NOTE ON THE USE OF THE WORDS "ACIDIC" AND "BASIC." Geologists have long distinguished between "acidic" and "basic" igneous rocks. The former are poor, the latter rich in soluble mineral salts. From the standpoint of the distribution of vegetation, the amount of soluble mineral salts in the soil is a fundamental matter. Acidic rocks, soils, peats, and waters are those which contain only a small amount of soluble mineral salts; and basic rocks, soils, peats, and waters are those which contain a large amount. In time, when more analyses have been made, it may be possible to express this relationship in quantitative terms; but, in the present state of knowledge, the matter can only be expressed in a general way. Acidic waters are neutral to acid in reaction, basic waters neutral to alkaline.

CHAPTER II

WOODLAND ASSOCIATIONS

Distribution of the woods. Woodland associations of Great Britain. Woodland associations of the southern Pennines. Factors related to the distribution of the woodland associations. Oak woods of *Quercus Robur*. Transitional woods of *Quercus Robur* and *Q. sessiliflora*. Oak woods of *Quercus sessiliflora*; trees and shrubs; variation of vegetation in the oak woods; influence of shade on the ground vegetation. Alder-willow thickets. Birch woods of *Betula pubescens*; the primitive birch forest. Ash woods of *Fraxinus excelsior*; semi-natural woods and plantations on the limestone slopes; trees and shrubs; herbaceous vegetation. Comparison of the woodland plants of the southern Pennines.

DISTRIBUTION OF THE WOODS

THE great majority of the woods occur on the slopes of the hills, where they ascend, on an average, to about 1000 feet (305 m.). Occasionally, they reach an altitude of 1250 feet (381 m.). On the Coal-measure rocks, which do not, as a rule, reach so high an elevation as the Pendleside, Millstone Grit, and Mountain Limestone rocks, the woods occasionally extend on to the flatter plateaux. In such situations, woods seldom occur at altitudes higher than 800 feet (244 m.). On the whole, the woods are of small size; and they by no means cover the whole of the hill slopes. In fact, as a rule, the hill-sides are occupied by uncultivated grassland (see Chapter IV) or scrub (see Chapter III), where they are not cultivated as permanent pasture (see Chapter VIII).

 W. B. Crump

Figure 3.

Oak Wood of *Quercus sessiliflora.*

The trees, being near their upper altitudinal limit and the soil being shallow, are small. The ground vegetation consists chiefly of the Bilberry (*Vaccinium Myrtillus*) on and around the boulders, and of the silver Hair-grass (*Deschampsia flexuosa*).

Woodland Associations of Great Britain

The various plant associations which are known to occur among British woodlands are summarised below (cf. Moss, Rankin, and Tansley, 1910):—

I. Alder and Willow Associations. *On very wet soils.*

A. *On soils supplied with acidic waters.*

1. Alder and willow thickets of lowland moors.

B. *On soils supplied with basic waters.*

2. Alder and willow thickets of the East Anglian fens.

C. *On fresh soils subject to periodical inundations.*

3. Alder and willow thickets by stream sides.

II. Oak and Birch Associations. *On non-calcareous soils.*

D. *On deep clays.*

4. Oak woods with *Quercus Robur* (= *Q. pedunculata*) dominant. Usually coppiced (see Moss, Rankin, and Tansley, 1910: 118). Widespread throughout the lowlands of southern and central England.

4 *b*. Oak-hornbeam woods with *Q. Robur* and *Carpinus Betulus* sharing dominance. Rather local in south-eastern England. Frequently coppiced.

E. *On dry sands and gravels.* Not uncommon in the south and east of England. Locally coppiced.

5. Oak woods with *Quercus Robur* or *Q. sessiliflora* dominant, either separately or in combination.

6. Birch woods with *Betula pubescens* dominant, or with *B. pubescens* and *B. alba* (= *B. verrucosa*) in combination.

7. Pine woods with *Pinus sylvestris* dominant.

7 *a*. Mixed woods of oaks, beeches, birches, and pines.

8. Beech woods with *Fagus sylvatica* dominant. Local in the south of England.

F. *On the shallow soils of the older siliceous rocks.*

9. Oak woods with *Q. sessiliflora* dominant. Very common on hill slopes in the north and west of Britain at altitudes below 1000 feet (305 m.). Locally coppiced.

10. Oak-ash woods with *Q. sessiliflora* and *Fraxinus excelsior* sharing dominance. Very local. They occur in the Lake District, where the rainfall is very high.

11. Birch woods with *Betula pubescens* dominant. Local. They occur in the north of England as a zone above the woods of *Quercus sessiliflora* at altitudes between 1000 feet and 1250 feet (381 m.). Much commoner in Scotland.

12. Birch-ash woods with *B. pubescens* and *Fraxinus excelsior* sharing dominance. Local. They occur in the Lake District.

13. Birch-pine woods with *B. alba; B. pubescens* and *Pinus sylvestris* var. *scottica* sharing dominance. Local and confined to mid-Scotland.
14. Pine woods with *Pinus sylvestris* var. *scottica* dominant. Local and confined to mid-Scotland.

III. ASH AND BEECH ASSOCIATIONS. *On calcareous soils.*

G. *On deep marls or calcareous clays.*

15. Ash-oak woods with *Fraxinus excelsior* and *Quercus Robur* sharing dominance. Abundant in the south of England. Usually coppiced.

H. *On shallow calcareous soils on hill slopes of the fissured limestone rocks.*

16. Ash woods with *Fraxinus excelsior* dominant. Frequent in the west and north of England up to about 1000 feet (305 m.), and local on the chalk.
17. Ash-birch woods with *Fraxinus excelsior* and *Betula pubescens* sharing dominance. Local on the upper slopes of hills of Carboniferous Limestone in the north of England.

I. *On shallow and very calcareous soils on chalk rock.*

18. Beech woods with *Fagus sylvatica* dominant. Frequent on the chalk escarpments in south-eastern England.

WOODLAND ASSOCIATIONS OF THE SOUTHERN PENNINES

Of the woodland associations above enumerated, two are well developed on the southern Pennines, and three others are only moderately well represented. The two former are the association of *Quercus sessiliflora* on damp, shallow, siliceous soils, and the association of *Fraxinus excelsior* on shallow calcareous soils; and these woods are described in some detail in this chapter. The woods which are only moderately well represented on the southern Pennines are the association of *Quercus Robur* on deep sandy soils, the association of *Betula pubescens* on damp, shallow, siliceous soils, and the alder-willow thickets of stream sides. Some transitional and intermediate woods also occur, and these will be referred to in the proper places.

The relationships of these associations may be conveniently set out in the following form:—

<table>
<tr><td rowspan="5">WOODS WITH</td><td>Alnus glutinosa and Salix spp. dominant</td><td></td><td>ALDER-WILLOW THICKETS</td></tr>
<tr><td>Quercus Robur dominant</td><td rowspan="2">Oak woods</td><td rowspan="3">OAK AND BIRCH WOODS</td></tr>
<tr><td>Q. sessiliflora dominant</td></tr>
<tr><td>Betula pubescens dominant</td><td>Birch woods</td></tr>
<tr><td>Fraxinus excelsior dominant</td><td></td><td>ASH WOODS</td></tr>
</table>

The distinction between the oak and birch woods on the one hand and the ash woods on the other hand is very sharp and clear in the Peak District, where most of the woods may without difficulty be referred to the oak and birch woods or to the ash association. It is true that a number of the woods are exploited for timber. However, in many of these, no re-planting takes place; and the indigenous trees spring up again quite spontaneously, either from the cut stools or from self-sown seed. In other cases, non-indigenous trees, such as beech (**Fagus sylvatica*[1]), sycamore (**Acer Pseudoplatanus*), larch (**Larix decidua*=**L. europaea*), and pine (**Pinus sylvestris*) are planted where the native trees have been felled; but, even in these cases, unless the shade cast by the planted species differs greatly from that cast by the original ones, the ground flora usually affords a fairly conclusive test as to whether or not the original wood belonged to the oak and birch woods or to the ash wood. A few of the woods of the district may indeed be said to be really primitive, as human interference with them is confined to the occasional cutting down of one or two trees by the occupier of some upland farm.

Many of the woods, however, are in a degenerate condition; and there is in this district no sharply dividing line between degenerate woodland on the one hand and scrub (considered in the next chapter) on the other. The questions relating to existing plantations and to reafforestation are discussed in the last chapter of the book.

Factors related to the Distribution of the Woodland Associations

Sufficient examples of woodland have now been examined in this and other districts to enable one to judge, in a general way at least, which are the principal ecological factors related to the present distribution of the various woodland associations. The oak and birch woods of the Pennines, as contrasted with the ash woods of the same region, are related to a difference in the chemical nature of the soil; for the former woods are here strictly confined to non-calcareous soils, and the latter, with

[1] Throughout this book, the species which are not indigenous are preceded by an asterisk.

equal strictness, to calcareous soils. The associations of *Quercus sessiliflora* and of *Betula pubescens*, as contrasted with the association of *Quercus Robur*, are also edaphic associations; for the former occur solely on the shallow siliceous soils of the sandstones and shales whilst the latter is limited to the deep fluvio-glacial sands.

On the other hand, the association of *Quercus sessiliflora*, as contrasted with that of *Betula pubescens*, is related to those climatic factors connected with altitude; for both associations occur on shallow siliceous soils, but the former occurs at altitudes below 1000 feet (305 m.) and the latter at altitudes between 1000 feet and 1250 feet (381 m.).

The mean annual rainfall of the sites of the various woodland associations varies locally from about 35 inches (89 cm.) to about 50 inches (127 cm.); but, so far as one is able to judge, there is, in this district, no definite relationship between the different woodland associations and local differences in rainfall. On the whole, it may perhaps be said that the woods of *Quercus Robur* have the lowest rainfall and those of *Betula pubescens* the highest rainfall; whilst the rainfall of the woods of *Quercus sessiliflora* and of *Fraxinus excelsior* is intermediate and approximately equal in amount.

As regards the water-content of the soils of the various woods, that of the woods of *Quercus Robur* is the lowest, that of the woods of *Q. sessiliflora* and *Fraxinus excelsior* is rather higher, varying from moderately dry to very wet, and that of the woods of *Betula pubescens* is, on the whole, the highest. The surface soils of the birch association and of the two oak associations show a marked tendency to form acidic humus, a tendency which is scarcely discernible in the soils of the ash woods. The nature of the surface soil is doubtless important in relation to the germination of the seeds of the dominant species, and therefore in relation to the rejuvenation of the woods; but few or no experiments appear to have been conducted with the view of testing such a hypothesis. In any case, the surface soil cannot have much to do with the biology of mature trees with deep tap roots, such as the ash.

Factors, then, which appear to be of importance in accounting for the differential distribution of the various woodland associations of the Peak District are (1) the chemical nature of the

Copyright *W. B. Crump*

Figure 4.

Oak Wood of *Quercus sessiliflora*.

Shrubby undergrowth. Dog Rose (*Rosa canina*) in foreground and Hazel (*Corylus Avellana*) behind.

soil, (2) the depth of the soil, and (3) the altitude of the woodland site. Other factors are of importance with regard to the distribution of the various types of ground vegetation (see pp. 53 and 71).

1. The chemical nature of the soil. The soils of the slopes of the limestone hills, to which the ash woods are limited, are rich in calcium carbonate: on newly exposed soils in such situations, the soil, in fact, may consist almost wholly of this substance; but commonly the percentage of lime varies from about five to about thirty per cent. The percentage of lime must be very high in the rock-strata in which the tap roots of the ash are fixed. On the chalk of the south-east of England, the woods are usually dominated by the beech (*Fagus sylvatica*), though the ash association does occur to some extent on the chalk.

The soils of the sandstones and shales have a very low lime-content. In several analyses, the percentage of lime (calculated as calcium oxide) was so low as 0·02 per cent.; and in no case was it higher than 0·05 per cent. On these soils, ash woods do not occur, whereas oak and birch associations are numerous and typical. It is not suggested here that the presence of lime is the direct cause of the occurrence of ash woods on the limestone slopes, or that the absence of it is the direct cause of the occurrence of oak and birch woods on the siliceous slopes, as much more experimental work must be done by plant physiologists before this classical problem can be approached from a point of view which is likely to afford a satisfactory outlook on the problems involved: all that is stated is that, so far as this district is concerned, the ash woods invariably occur on the calcareous strata and never elsewhere, and that the oak and birch woods invariably occur on the siliceous rocks and never on the calcareous ones. It seems, however, reasonable to suggest that the presence or absence of a high lime-content of the soil is concerned either directly or indirectly with the present distribution of the principal types of woodland of this district.

2. The depth of the soil. The fluvio-glacial sands in the west of the district are deep, and hence offer no physical obstacle to the growth of the long tap roots of *Quercus Robur*. Woods of *Quercus Robur* occur on these soils. Further, so far as this district is concerned, such woods are restricted to these soils.

The woods of *Quercus sessiliflora* and of *Betula pubescens* are confined to the sandstones and shales. The soil over the sandstones is shallow, that is, shallower than would appear to be necessary for the free growth of the tap roots of *Quercus Robur*; and similarly the soil of the shales is frequently shallow, being often interrupted by bands of flagstone and half-formed sandstone. The restriction of woods of *Quercus sessiliflora* to shallow, siliceous soils is very general throughout the whole of the British Isles; but exceptional cases occur on sandy and gravelly soils in the south-east of England.

3. The altitude of the woodland site. Woods of *Quercus sessiliflora* cease at altitudes of about 1000 feet (305 m.), whilst woods of *Betula pubescens* reach altitudes of about 1250 feet (381 m.). As the physical and chemical conditions of the soils of the two associations are identical, it seems clear that the failure of the oak woods to reach the higher altitude is due to the severer climatic conditions which prevail there. Such conditions are the lower temperatures, the greater velocity of the wind, and (to some extent) the lower atmospheric humidity and higher rainfall.

OAK WOODS OF *QUERCUS ROBUR*

In this district, woods of the peduncled oak (*Quercus Robur*) occur only on the western boundary; and they represent the eastern fringe of the lowland oak woods of the Cheshire plain. The latter, in their turn, are a north-western continuation of the lowland oak woods which are characteristic of the Midlands and of the south of England; and an eastern extension of the latter may be traced through Nottinghamshire and into east Yorkshire. As already indicated, oak woods of *Quercus Robur* may occur either on deep clays or on deep sands or gravels. Woods of both these types occur on the Cheshire plain, though the oak woods of the sandy and gravelly soils are there more extensive than those on clayey soils. It is the woods of the former type that just reach the western margin of this district.

The best examples of woods of *Quercus Robur* in this district occur in the lower courses of the rivers Etherow and Goyt, two of the head-streams of the river Mersey. Ernocroft Wood and Bottoms Hall Wood are examples of such woods, where *Quercus Robur* is the dominant tree and where *Q. sessiliflora* is rare. The whole of this area has been mapped by the geological surveyors as consisting of rocks of the Coal-measure series; but such rocks are here obscured, or partially obscured, by deep deposits of fluvio-glacial sands. There are, unfortunately, no geological maps with drift published of any portion of the district; and, in fact, the Geological Survey has not yet commenced its examination of the drift of this locality. When, however, such maps are published and compared with vegetation maps, they will afford a striking commentary on the differential distribution of the woods of *Quercus Robur* and *Q. sessiliflora* on the west of the Pennines in this latitude.

The region of woods of *Quercus Robur* of the lowlands of the east of the Pennines does not reach the Peak District, and, in fact, possibly ceases east of Sheffield.

The woods of *Quercus Robur* of this district occur at their local, upper altitudinal limit, and have been interfered with by the planting of foreign trees and shrubs, such as laurels and rhododendrons; and they cannot therefore be regarded as typical of such woods in general. Hence it is not desirable to describe them here in detail. Such a description will be a more fitting outcome of a vegetation survey of some portion of the Cheshire plain, where woods of *Quercus Robur* are more extensive, more numerous, and more typical in character.

TRANSITIONAL WOODS OF *QUERCUS ROBUR* AND *Q. SESSILIFLORA*

In a few places near the junction of the woods of *Quercus Robur* and of *Q. sessiliflora*, some small woods occur in which the two species are found side by side. This is the case, for example, with regard to the small Townscliffe Wood, east of Mellor. Had such woods been extensive enough, their transitional character could have been indicated on the vegetation maps by giving them the ground colour of the woods of *Quercus Robur* and stippling on this the darker colour used for woods of *Q. sessiliflora*. However, this course was impracticable owing to the small size of the transitional woods.

Whilst *Quercus Robur* and *Q. sessiliflora* respectively form, as a rule, well-defined associations, it sometimes happens that a stable association occurs in which the two species are present in almost equal proportions. Such transitional woods may occur in localities where the two associations come into close propinquity, and also sometimes on dry sandy or gravelly soils. In such transitional woods, the hybrid oak (*Quercus Robur* × *sessiliflora*) invariably occurs. The occurrence of this hybrid, though quite general in such situations, was unsuspected, so far as this country is concerned, until quite recently (see Moss, 1910 *a*: 34).

OAK WOODS OF *QUERCUS SESSILIFLORA*

Woods of *Quercus sessiliflora* occur on the damp, shallow, siliceous soils of the Coal-measures, the Millstone Grit, and the Pendleside (or Yoredale) rocks, up to an altitude of about 1000 feet (305 m.). The rocks of the Coal-measure series, as a rule, occur at moderately low altitudes, flanking the Pennine watershed. On these rocks, the woods, whilst mainly confined to the slopes of the hills, occasionally extend some little distance on to the plateaux, whereas on the rocks of the Millstone Grit and the Pendleside series, the woods are almost entirely confined to the steep slopes of the narrow valleys or "cloughs"; and it is only rarely that they spread out on to the flatter and more exposed plateaux.

The woods of *Quercus sessiliflora* of the southern Pennines have been described by previous writers. Crump (1904: xxxiii) subdivided them into (*a*) mixed deciduous woods, and (*b*) dry oak woods on the Coal-measure, Millstone Grit, and Yoredale (or Pendleside) rocks. Smith and Moss (1903: 387) and Smith and Rankin (1903: 159) adopted almost the same subdivisions when they described (*a*) lowland oak woods and (*b*) upland oak woods. Woodhead (1906), in describing their ground vegetation, referred to them as (*a*) mixed deciduous woods of the Coal-measure area (p. 336), and (*b*) mixed deciduous woods of the Millstone Grit area (p. 347). However, all these subdivisions are only particular aspects of the association of *Quercus sessiliflora* as developed on shallow, siliceous soils. The "pine woods" of Smith and Rankin (*loc. cit.*) are merely plantations of conifers, on sites previously occupied by woodland, grassland, or farmland.

Copyright *W. B. Crump*

Figure 5.

Oak Wood of *Quercus sessiliflora*.

The ground vegetation is composed largely of the Bluebell (*Scilla non-scripta*).

Trees and Shrubs

The sessile-fruited oak (*Quercus sessiliflora*) is undisputably the prevailing tree of these woods. Because of the comparatively great amount of light which penetrates the oak canopy, the ground is fully covered by vegetation; and therefore the tree may be spoken of as the dominant species, as it exercises a controlling influence on the rest of the vegetation of the wood. In the earlier accounts (*op. cit.*) of the vegetation of the Pennines, the plant was referred to as "*Quercus Robur* Linn." It has, however, been shown (Moss, 1910 *a*) that this name refers to the peduncled oak alone, and that it is possible and desirable to distinguish separate associations of the two British species (*Q. Robur* and *Q. sessiliflora*) of oak. In this district, *Quercus sessiliflora* grows well and forms moderately large trees up to an altitude of about 800 feet (244 m.), particularly on the shales: above this altitude, especially on soils over the massive sandstone rocks, the trees are apt to be of short stature and of small girth; and near the present altitudinal limit of woodland, the trees are often little bigger than shrubs (see figure 3). Seedlings are often met with in the damper woods, but are rare in those with a peaty soil at the higher altitudes.

The pedunculate oak (**Quercus Robur*) is absent from the great majority of the oak woods of the Pennine slopes. As has been already stated, this species is found in a few of the western woods situated on the glacial sands: elsewhere in the district it only occurs as a planted tree, along with other aliens, such as the sycamore, the beech, the larch, and the pine. It is present, as might be expected, in most of the newer plantations and in park-lands; but, even in such localities, it is not very abundant, and it rarely grows to a large size.

No conifers are indigenous in the woods of *Quercus sessiliflora* of this district; but the Scots pine (**Pinus sylvestris*) and the larch (**Larix decidua*) are frequently planted. Other conifers occasionally or rarely met with in the woods are the black or Austrian pine (**Pinus austriaca*), the spruce fir (**Abies excelsa*), and the Douglas fir (**Pseudotsuga Douglasii*).

As sub-fossil timber, the Scots pine is occasionally found buried under the peat of the southern Pennines; and it is rather remarkable therefore that the plant is not indigenous in the Pennine woods at the present time. Still, judging from the paucity of the number of the records of buried pine timber, the species does not appear to have been more than an occasional or, at most, a locally abundant integer even in the prehistoric woods of the Pennines; and it probably became extinct at a very remote date. At the present time, seedling pines are not abundant on the Pennines even in and around pine plantations; and the tree does not flourish nearly so well on these hills as on the dry and sandy heaths in the south of England.

The only species of poplar which is indigenous in this district is the aspen (*Populus tremula*); and even this species appears to be quite rare in the oak and birch woods. Other poplars (*e.g.*, **P. canadensis* and **P. candicans*) are planted occasionally, though more frequently on the outskirts of the woods than inside the woods themselves.

Several indigenous species of willow occur. *Salix cinerea* is common, and ascends to nearly 1200 feet (366 m.) in some of the cloughs. *S. caprea* is occasional; but both *S. cinerea* and *S. caprea* are absent from the driest woods. *S. aurita* is local, but occasionally forms thickets in damp spots in the cloughs. *S. pentandra* is also rare. Hybrids of *S. caprea*, *S. cinerea*, and *S. aurita* are not uncommon. *S. repens* occurs, but is rare. *S. fragilis*, *S. viminalis*, *S. caprea* × *viminalis* occur locally by the stream sides, at altitudes below 600 feet (183 m.). Although *S. alba*, *S. purpurea*, and × *S. rubra* (= *S. purpurea* × *viminalis*) are recorded (Linton, 1903), they are perhaps not indigenous in the Peak District.

The hazel (*Corylus Avellana*) is rather abundant in the damper woods, but much rarer in the drier ones.

The common birch (*Betula pubescens*) is, on the whole, the most constant and the most abundant associate of the sessile oak. Forms or varieties with glabrous or sub-glabrous twigs (*B. pubescens* forma *denudata*) are not uncommon: *B. pubescens* var. *parvifolia* is rare, but has been observed. In some of the woods, however, the birch is rare or absent. Not infrequently, the birch becomes locally dominant in places where extensive felling of the oak and no subsequent planting, have taken place; and its small, light, and winged fruits are evidently of more advantage in colonizing cleared or partially cleared areas than the large and heavy fruits of the oak. In some such places, seedlings of the birch are extremely abundant; and local plant societies of well-grown birches are common in many of the oak woods. A birch wood at low altitudes in this district usually represents a degenerate oak wood which has been colonized by birches. In ascending the cloughs, many of which are now almost treeless, isolated plants of the common birch are often the last trees which are encountered. Under the peat of the moors, birch remains are locally very abundant.

The white birch (**Betula alba*) is perhaps not indigenous in the hilly woods of the Peak District, as it only seems to occur in the company of such obviously planted trees as the beech, sycamore, peduncled oak, larch and pine. It is never abundant; and it is absent over extensive tracts and from the more primitive and the more upland woods.

The alder (*Alnus glutinosa*) is confined to stream sides and marshy places, where it often forms small societies. It is more abundant in the oak woods than in the ash woods.

The beech (**Fagus sylvatica*), although an almost invariable constituent of the larger woods, has little claim to rank as indigenous. As a rule, evidences of its introduction are easy to trace, either because it occurs in obviously recent plantations or because historical evidence of planting is procurable. Whether indigenous or not, the beech grows well on all the Pennine slopes, both siliceous and calcareous. In favourable seasons, ripe fruits have been observed on trees at an altitude of 1500 feet (457 m.). First-year seedlings are frequently met with in the woods: older seedlings, however, have not been observed; and there is no evidence to show that the beech rejuvenates itself in these hilly woods. Apparently the seedlings of the beech are all destroyed during their first winter, perhaps because they are unable to endure the alternating cold and mild periods characteristic of the markedly insular climate of the north and west of the British Isles. On the other hand, the rejuvenation of the beech may be observed on sandy and on chalky soils in the south of England, where the tree is indigenous. The analogy of the beech to the pine in the matter of rejuvenation in the north and south of England is remarkable; and it may be that the pine, as well as the beech, is indigenous in the south-east of England. As a planted tree, the beech is locally abundant up to about 1600 feet (488 m.).

The Spanish or sweet chestnut (**Castanea sativa* = **C. vulgaris* = **C. vesca*) is planted rarely up to 1500 feet (457 m.); but it is seldom a success in this hilly district; and its fruits do not ripen on the Pennines. In some of the lowland oak woods of Cheshire, as in Delamere Forest, the tree is much more successful; and the tree is said to ripen its fruits occasionally in one or two localities of that county. On sandy soils in the south and east of England, the tree not uncommonly ripens its fruits, as in Kent, Bedfordshire, and Cambridgeshire; and there young trees and seedlings may be seen in all stages of development in certain woods and plantations.

The wych elm (*Ulmus glabra* = *U. montana*) is indigenous, and occurs up to about 1000 feet (305 m.). Above this altitude, it is frequent in plantations up to 1500 feet (457 m.). It is a constant and sometimes an abundant constituent in the damper woods, but is rare in the drier ones. In favourable localities, seedlings are very common. The seeds germinate very shortly after they fall from the tree; and seedlings may be found in August on damp, bare soil in sheltered situations.

The hawthorn (*Crataegus Oxyacantha* = *Q. monogyna*) is an occasional associate in the damper and more shady woods, and an abundant one in the drier and more exposed woods. Frequently, it is the last relic of pre-existing woods on exposed hill-sides. The form or variety *laciniata* is common; and this indeed may be the indigenous form.

The crab apple (*Pyrus Malus*) is never more than a shrub on the hills of northern England. The remark in Linton's flora (1903: 142) that it is "common everywhere" in Derbyshire is a curious over-statement. In the woods of *Quercus sessiliflora* the plant is rather local, and rarely, if ever, abundant. In many seasons, it fails to ripen its fruits.

The rowan or mountain ash (*Pyrus Aucuparia*) occurs in most of the woods; and, in rocky, upland, and heathy situations, it is often abundant. It is frequently the last isolated tree seen in ascending the cloughs.

The raspberry (*Rubus Idaeus*) is abundant locally, preferring damp soils without much acidic humus. *R. fissus*

Copyright *W. B. Crump*

Figure 6.

Oak Wood of *Quercus sessiliflora.*

The ground vegetation consists chiefly of the Bracken (*Pteris aquilina*) and the wood Soft-grass (*Holcus mollis*).

occasionally occurs in similar situations. The bramble or blackberry is very abundant, "*Rubus Selmeri* and *Rubus dasyphyllus* reaching the highest altitude" (Linton, 1903: 114) of any of the segregates of this polymorphous group. The dewberry (*R. caesius*) is confined to the lower altitudes, where it is rather local.

Briers or wild roses (*Rosa* spp.) are common and generally distributed, especially *R. canina*. *R. tomentosa* is local; and "*R. mollis*" has been recorded. *R. arvensis* is locally abundant at the lower altitudes. Other species of Rosa and Rubus are enumerated at the end of the chapter.

Sloe or blackthorn (*Prunus spinosa*) is a constant, though, as a rule, an infrequent member of the association. Rarely, as by the stream sides in some of the cloughs, it forms dense thickets. It rarely ripens its fruits at altitudes greater than 600 feet (183 m.). Bird-cherry (*Prunus Padus*) is abundant in some of the cloughs; but it becomes more and more uncommon as the plains are approached. Conversely, the cherry (*Prunus avium*) is commoner at the lower levels, and fails to ascend higher than about 600 feet (183 m.).

Gorse or furze (*Ulex europaeus*) is thinly scattered through the woods at the lower altitudes; and the dwarf furze (*Ulex Gallii*) is often abundant on the outskirts of the woods at the higher altitudes.

Broom (*Cytisus scoparius*) is found but rarely inside the woods; but, like the two species of Ulex, it is often abundant on the outskirts of the woods.

The holly (*Ilex Aquifolium*) occurs in almost every oak wood of the district, and is typically abundant where the soil is moderately dry. It is occasionally the last isolated tree seen in ascending the cloughs. Linton (1903: 97) records it as occurring at an altitude of 1050 feet (318 m.) in Jagger's Clough. The plant rarely produces flowers, and still more rarely produces ripe fruit at the higher altitudes on the Pennines.

The maple (*Acer campestre*), as a shrub, is confined to comparatively low altitudes, and is not encountered at all in the higher and remoter cloughs. As a tree, it is almost if not quite unknown in the woods of the district. The sycamore (**Acer Pseudoplatanus*) is always an introduction, though, as a planted tree, it is locally abundant and succeeds well. First-year seedlings of sycamore are common; and occasionally these grow up into trees.

No species of lime is indigenous in the oak and birch woods with the possible exception of the small-leaved species (*Tilia cordata* = *T. ulmifolia* = *T. parviflora*); and even this is extremely rare. The common lime (*T. europaea* = *T. vulgaris*) is never a woodland plant; and the broad-leaved lime (*T. platyphylla* = *T. grandifolia*) is not indigenous in Derbyshire.

Ivy (*Hedera Helix*) is a constant and frequent associate in all the woods.

The ling or heather (*Calluna vulgaris*) is confined to woods whose soil contains much acidic humus or peat, and to the more open parts of such woods.

The bilberry (*Vaccinium Myrtillus*) occurs in somewhat similar situations to the heather, and usually grows with it; but it also thrives in more shady parts of the wood than that plant (see figure 7). The cowberry (*V. Vitis-idaea*) is local, but very abundant in some degenerate woods with a peaty soil, as in Longdendale, near Crowden railway station.

The ash (*Fraxinus excelsior*) is, in the oak woods, almost limited to stream sides and swampy places. In the drier oak woods it is very rare.

The elder (*Sambucus nigra*) is locally abundant in the lower woods, but rather uncommon at the higher altitudes. In places where the woods are disturbed and especially near villages, the plant frequently overruns the ground.

The guelder rose (*Viburnum Opulus*) is, in all the damper woods, an occasional associate, and reaches comparatively high altitudes (about 318 m.).

The honeysuckle (*Lonicera Periclymenum*) is abundant and general. This plant and the ivy are the only indigenous lianes of the north of England.

Variation of Vegetation in the Oak Woods

The vegetation of the oak woods varies greatly from place to place. The variation in the vegetation is due to the variation of the various ecological factors. It is impossible, in the present state of knowledge, to give anything like a complete account of these factors; and it is still more difficult to state the action of the various factors either on the vegetation as a whole or upon the individual plants. However, some operating factors may be recognised; and doubtless future work by plant physiologists will suggest what are the effects of these factors on the vegetation and on the individual plants.

In the woods of *Quercus sessiliflora*, important ecological factors are the water-content of the soil, the kind and quantity of humus present, and the amount of light which penetrates the leaf-canopy of the trees and shrubs. These factors are correlated in the most complex manner, and can perhaps best be illustrated by considering various woodland habitats where any one of them becomes pronounced. It must be remembered, however, that the habitats chosen are connected by all possible intermediate stages; and it is the sum of these minor habitats which constitute the more general habitat of the association as a whole.

(1) Marshy places. Where springs arise and by the sides of the various rills and streams, wet and marshy places occur where the soil is well aërated; and consequently any humus that is present is mild (not alkaline) humus and not acidic humus. In such places, the oak (*Quercus sessiliflora*) tends to become very rare, and the alder (*Alnus glutinosa*) and the ash (*Fraxinus excelsior*) to become correspondingly more abundant. Birches (*Betula pubescens*, and *B. pubescens* var. *parvifolia*) may however remain. At the lower altitudes, the crack willow (*Salix fragilis*) is sometimes found; and at the higher altitudes, where, however, the soil-water may be more or less acid, *Salix aurita* is locally abundant. *S. cinerea* is usually an abundant species; and where the two last are found together, hybrids (*S. aurita* × *cinerea*) occur. The bird

cherry (*Prunus Padus*) is locally abundant. The ground flora is often characterised by tufts of the common rush (*Juncus effusus*); and a smaller species of rush (*J. lamprocarpus*) is often strongly in evidence. Many marsh plants occur, such as the meadow sweet (*Spiraea Ulmaria*) and the great Valerian (*Valeriana sambucifolia*). Ferns are abundant, particularly the lady fern (*Athyrium Filix-foemina*), and also the wood horsetail (*Equisetum sylvaticum* and *E. sylvaticum* var. *capillare*). Almost any of the plants which ordinarily occur in a Juncus swamp (see page 147) may be found; whilst the following additional species, although very characteristic, are more or less local in their occurrence:—

Phegopteris Dryopteris (rare)
P. polypodioides (local)
Nephrodium montanum
(=N. Oreopteris)
Athyrium Filix-foemina
A. Filix-foemina *var.* rhoeticum
Rumex Acetosa
Stellaria nemorum (local)
Trollius europaeus (rare)
Ranunculus Ficaria
Cardamine amara (local)
C. flexuosa
Chrysosplenium oppositifolium
C. alternifolium (local)
Spiraea Ulmaria
Geum rivale
G. rivale × urbanum (rare)
Geranium sylvaticum (rare)
Angelica sylvestris
Lysimachia vulgaris
Myosotis palustris
Mentha aquatica
Solanum Dulcamara
Scrophularia nodosa
Valeriana dioica (rare)
V. sambucifolia
Petasites ovatus
Cnicus palustris
C. heterophyllus (rare)
Crepis paludosa (local)
Deschampsia caespitosa
Carex remota (rare)
C. pendula (rare)
C. sylvatica (rare)
C. helodes
(=C. laevigata)
Luzula maxima
Juncus effusus

(2) Damp places with mild humus. It is in these places that one finds the most characteristic "woodland plants"; but, in a hilly district like this, shade-loving species and their typical habitat are much more local in occurrence than in lowland woods. Here they occur, as a rule, on moderate and well-drained slopes, usually near streams, and where the light is not too strong. The oak is the dominant tree: the wych elm occurs rather abundantly, and the wild cherry (*Prunus Cerasus* and *P. avium*) rather rarely. Birches and holly are typically absent. Shrubs are moderately abundant, especially

Copyright *W. B. Crump*

Figure 7.

Oak Wood of *Quercus sessiliflora.*

The ground vegetation consists largely of the Bracken (*Pteris aquilina*), the Bilberry (*Vaccinium Myrtillus*), and the silver Hair-grass (*Deschampsia flexuosa*).

hazel, sallows, maple, wild roses, ivy, and honeysuckle. The following herbaceous species are characteristic:—

Nephrodium Filix-mas	Myosotis sylvatica (local)
N. spinulosum	Ajuga reptans
N. dilatatum	Prunella vulgaris
Aspidium aculeatum (rare)	Lamium Galeobdolon
A. angulare (rare)	Stachys sylvatica
Athyrium Filix-foemina	Veronica montana
Rumex Acetosa	Asperula odorata
Lychnis dioica	Campanula latifolia (rare)
(=Melandrium rubrum)	Lactuca muralis (local)
Stellaria Holostea	Hieracium vulgatum (local)
Aquilegia vulgaris (v. rare)	H. boreale (local)
Anemone nemorosa	Milium effusum (local)
Ranunculus auricomus	Poa nemoralis (rare)
Fragaria vesca	Melica uniflora
Geum urbanum	Festuca gigantea
Vicia sepium	Bromus ramosus
V. sylvatica (v. rare)	Arum maculatum
Geranium Robertianum	Allium ursinum
Oxalis Acetosella	Scilla non-scripta
Mercurialis perennis	Paris quadrifolia (v. rare)
Viola Riviniana, *var.*	Gagea lutea (v. rare)
Epilobium montanum	Narcissus Pseudo-Narcissus (local)
Circaea lutetiana	Tamus communis
Sanicula europaea	Neottia Nidus-avis (v. rare)
Conopodium majus	Listera ovata (local)
Heracleum Sphondylium	Helleborine latifolia (rare)
Primula vulgaris (rare)	Orchis mascula (local)

(3) Dry places with some acidic humus in the upper layers of the soil. The oak is dominant: birches and holly are usually present: the wych elm tends to be uncommon or rare: shrubs, such as hazel and *Salix capraea*, are fairly abundant: ivy, honeysuckle, roses, and brambles tend to be abundant. The ground species with showy flowers are of local occurrence, except the bluebell which is often very abundant, growing in great masses (see figure 5), especially if the soil is not too dry. The typical ground vegetation is that described by Woodhead (1906: 344) as a "meso-Pteridetum," in which the bracken (*Pteris aquilina*), the bluebell (*Scilla non-scripta*), and the wood soft-grass (*Holcus mollis*) occur as social plants (see figure 6). The shade cast by the trees and shrubs is less than in the preceding type of

ground vegetation. The following ground species may also be expected to occur:—

Nephrodium dilatatum	Digitalis purpurea
N. Filix-mas (local)	Galium saxatile
Pteris aquilina	Campanula rotundifolia
Rumex Acetosella	Hieracium boreale
Potentilla erecta (local)	Holcus mollis
Arenaria trinervia	Deschampsia flexuosa (local)
Polygala serpyllacea	Anthoxanthum odoratum
Viola Riviniana *var.* diversa	Poa trivialis (local)
Conopodium majus	Carex pilulifera (local)
Teucrium Scorodonia	Luzula pilosa
Lamium Galeobdolon	Scilla non-scripta

(4) Places where acidic humus is present in good quantity. The oak remains dominant: the birch and the mountain ash are often abundant: the wych elm and the ash are rare or absent: shrubs are rather rare and sometimes almost absent; whilst heathy dwarf shrubs, especially the bilberry, are typically abundant (see figures 3 and 7). Ground species with showy flowers are rare: the soft-grass (*Holcus mollis*) is local and often absent; but the silver hair-grass (*Deschampsia flexuosa*) becomes exceedingly abundant: the bracken varies from being very abundant to very rare. This type of vegetation was termed by Woodhead (1906: 347) a "xero-Pteridetum." It is characterized by the following species:—

Nephrodium dilatatum	Digitalis purpurea
Blechnum spicant	Melampyrum pratense
Pteris aquilina	*var.* montanum
Potentilla erecta	*var.* hians
Ulex Gallii (local)	Galium saxatile
Cytisus scoparius (local)	Solidago Virgaurea
Lathyrus montanus	Hieracium vulgatum (local)
Polygala serpyllacea	Holcus mollis (local)
Pyrola media (v. rare)	Deschampsia flexuosa
P. minor (rare)	Molinia caerulea (local)
Calluna vulgaris	Carex binervis
Vaccinium Myrtillus	C. pilulifera
V. Vitis-idaea (local)	Luzula pilosa
Teucrium Scorodonia	L. multiflora

The two last-mentioned types of ground vegetation occupy by far the major portion of the woodland area, and they have been described in some detail by Woodhead (1906: 336, *et seq.*). In the case of a particular wood near Huddersfield, Woodhead showed by means of sketch maps that under trees like the oak and the birch, which admit a large amount of light, the bracken flourishes; but under trees with a much closer canopy, such as the wych elm, sycamore [introduced], and beech [introduced], much light is cut off, and the bracken is almost or entirely absent. Woodhead applied the same method to another plant, the bluebell (*Scilla non-scripta*), in the same wood. In this case, he found that light is not the [only] controlling factor, but that the plant is most abundant in a lighter, loamy soil where more shade occurred. The bluebell also occurs in stiffer soils simulating clay, and here competes with the yellow dead-nettle (*Lamium Galeobdolon*), the dog's mercury (*Mercurialis perennis*), the cuckoo-pint (*Arum maculatum*), and numerous root-branches of small trees. In shallow, sandy soil mixed with humus, the bluebell only occurs in straggling patches. On such soils, the bluebell does not form those unbroken stretches so characteristic of moist areas (cf. figure 5). The grassy vegetation of many of these woods is mainly composed of *Holcus mollis* and *Deschampsia flexuosa.* The latter species prefers an acidic, humous soil, where it is associated with *Galium saxatile, Vaccinium Myrtillus, Teucrium Scorodonia*, and *Solidago Virgaurea.*

The plant society in which the bracken (*Pteris aquilina*) is associated with *Holcus mollis*, Woodhead (*loc. cit.*) termed a "meso-Pteridetum," and that in which the bracken is associated with *Deschampsia flexuosa* he termed a "xero-Pteridetum." These terms, however, are not very satisfactory. First, the terms imply that the water-content of the so-called meso-Pteridetum is higher than that of the so-called xero-Pteridetum; but this is not established. Secondly, they imply that the members of the former association are "mesophytes" and those of the latter association "xerophytes"; but many characteristic members of the xero-Pteridetum can scarcely be admitted to the "ecological class" of xerophytes. Lastly, it is questionable if Schouw's termination *-etum* should be applied to any vegetation unit other than a plant association; and it is doubtful if the plant communities in question should be given a higher rank than that of ground societies.

Influence of Shade on the Ground Vegetation

There are practically no places within the oak woods where the shade is too great for the growth of a carpet of vegetation; but where the beech (**Fagus sylvatica*) has been introduced and planted in mass or even only in patches, such places occur. Where the shade is most dense under the beeches, practically no plants are to be found except a few mosses, and in autumn some of the larger fungi. Where the shade is less dense, a few grasses, such as *Holcus mollis* and *Deschampsia flexuosa*, may be scattered about; and mixed with them are a few shade-loving species, such as dog's mercury (*Mercurialis perennis*) and wood sanicle (*Sanicula europaea*).

The sycamore (**Acer Pseudoplatanus*) also casts much shade; but, as this is usually introduced singly among the indigenous oaks, the tree does not usually bring about a great change in the ground vegetation.

The pine (**Pinus sylvestris*) is frequently planted in mass; and, when planted closely, much shade is thrown, and the original ground species tend to disappear.

Of indigenous trees, the wych elm (*U. glabra = U. montana*) casts the greatest shade; but on the siliceous soils this species does not usually occur as a social tree except where planted.

ALDER-WILLOW THICKETS

In some of the valleys, there is, at the present time, no extensive tract of woodland at all. This is the case, for example, in the Edale Valley and around Chapel-en-le-Frith. There can be little doubt that, in all such cases, the primaeval woodland has been destroyed and the land put under cultivation. Trees, however, often line the banks of the streams which flow through such localities, and form narrow fringing thickets which may be a mile or so in length. The most abundant trees are the alder (*Alnus glutinosa*), willows (e.g. *Salix fragilis* and *S. cinerea*), and the ash (*Fraxinus excelsior*). These alder-willow thickets are to be regarded as the persisting and lingering remains of the alder-willow association which doubtless fringed practically all the streams when the latter flowed through the primaeval oak (*Quercus sessiliflora*) forest. On the accompanying vegetation maps, they are given the same colour as the oak woods of which they originally formed a part.

Most of the alder-willow thickets occur on the Pendleside shales; and on these soils at the lower altitudes, woods are now rare. It seems highly probable that this is due to the fact that the shales make excellent agricultural land; and the original woods on the shales, therefore, have nearly all been felled, and the land put down to cultivation, chiefly as permanent pasture (see Chapter VIII).

The flora of the alder-willow thickets does not differ materially from that of the damper parts of the oak woods. Occurring in the cultivated area, it is natural to find in them some alien trees, and some invading pasture species among the ground flora. The following trees and shrubs were noted in the ash-alder thickets near Edale and Castleton:—

Subdominant species

Salix fragilis	Alnus glutinosa
S. cinerea	Fraxinus excelsior

Locally abundant species

Salix caprea	Rubus spp.
Corylus Avellana	Rosa tomentosa (rare)
Betula pubescens	R. canina
Ulmus glabra	Pyrus Aucuparia
(= U. montana)	Hedera Helix
Prunus spinosa	Lonicera Periclymenum

Occasional and rare species

*Larix decidua	Q. sessiliflora
*Populus canadensis	*Prunus insititia
*Castanea sativa	Rosa arvensis
*Quercus Robur	*Acer Pseudoplatanus

Birch Woods of *Betula pubescens*

Birch woods were recognized in Scotland by Robert Smith (1900, *a* and *b*) who stated that they were quite natural and self-sown. Smith did not state which of the two British species form the dominant element; but both are actually abundant in Perthshire, though *Betula pubescens* ascends to higher altitudes than *B. alba*. In the Leeds and Halifax district, which lies immediately to the north of the Peak District, "the birch wood or uppermost tree zone of the Scottish Highlands is represented by a modification of the oak wood....The oak [*Quercus sessiliflora*] is usually dominant; but when the best of these are removed and no others planted, the birch [*Betula pubescens*] becomes dominant, either alone or with stunted oaks" (Smith and Moss, 1903: 388). In the Harrogate and Skipton district of the mid-Pennines, an uppermost woodland zone was also recognized where the dominant plants form a loose scrub of birch [*Betula pubescens*], mountain ash [*Pyrus Aucuparia*], holly [*Ilex Aquifolium*], hawthorn [*Crataegus Oxyacantha*], blackthorn [*Prunus spinosa*], and willows [*S. cinerea*, *S. caprea*, and *S. aurita*] (Smith and Rankin, 1903: 159).

Birch woods (see figure 8) are only feebly developed in the Peak District: none is of great extent; and none shows a sharp line of demarcation from the upper oak woods. As one ascends a wooded hill-slope composed of non-calcareous rocks, the oak (*Quercus sessiliflora*) becomes rare at altitudes above 1000 feet (305 m.), and usually ceases at 1100 feet (335 m.) or 1200 feet (366 m.). The diminution in number of the oaks is attended by an increase in number of the birches; so that there is a gradual transition from oak woods to oak-birch woods and to pure birch woods. Woods of the intermediate oak-birch type are of frequent occurrence in Longdendale and upper Derwentdale. Pure birch woods, however, are rare in the Peak District, though two or three rather small examples occur at altitudes above 1000 feet in the two valleys just mentioned. Along with the oak, most of the other trees, such, for example, as the alder (*Alnus glutinosa*), the wych elm (*Ulmus glabra* = *U. montana*), and the ash (*Fraxinus excelsior*), are left behind when the higher altitudes are reached. The mountain ash (*Pyrus Aucuparia*), on the other hand, becomes more abundant. Similarly, most of the shrubs of the oak woods become rare at the higher altitudes, the hawthorn (*Crataegus Oxyacantha*) and *Salix cinerea* perhaps ascending higher than most of the others. The number of shade-loving ground species also becomes greatly reduced; and such species (see page 55) are either totally absent or present in greatly

Copyright *W. B. Crump*

Figure 8.

Birch Wood of *Betula pubescens.*

The ground is marshy, and there is a great deal of the common Rush (*Juncus effusus*) and of the tufted Hair-grass (*Deschampsia caespitosa*).

reduced numbers. Since no Alpine and very few sub-Alpine species take the place of the absent lowland species, the floristic features separating the birch woods from the oak woods are, in this district, largely of a negative character. From the standpoint of vegetation, however, there are positive differences, as there is a rearrangement of the common members of the two associations.

There is little doubt that the birch woods of *Betula pubescens* must be placed in the same plant formation as the oak woods of *Quercus sessiliflora*, not merely because the two associations are connected by all possible gradations and because one may easily replace the other, but because the general habitats have so much in common and the floristic elements are so very much alike. A birch wood, in this district, is simply a wood in which the oaks, on account of the increased exposure consequent on the increased altitude, have largely or entirely disappeared, and in which the birches have profited by the absence of the competition of the oaks. However, the difference in altitude and the consequent differences of the flora and of the vegetation are sufficient to justify the placing of the two communities in separate associations, in spite of the existence of numerous connecting links.

Following a system of universal nomenclature (see Moss, 1910 *b*: 41, *et seq.*), the two most important woodland associations of the non-calcareous soils may be designated as follows:—

(i) Quercetum sessiliflorae or association of *Quercus sessiliflora*.
(ii) Betuletum pubescentis or association of *Betula pubescens*.

Regarding these as belonging to the formation Silicion, the above names may be combined as follows (Moss, *loc. cit.*):—

(i) Silicion Querceti-sessiliflorae.
(ii) Silicion Betuleti-pubescentis.

Not only do the meagre birch woods of the Pennines closely resemble the oak woods, but the more typical birch woods of Perthshire would appear to be very closely allied to the Scottish oak woods. R. Smith (1900 *b*: 45) in describing the birch woods of Perthshire says:—"The birch woods in the shelter of the river valleys may, however, have as rich a vegetation as the oak coppice. Thus, for example, a list taken in the birch wood between Loch Tummel and the Falls of

Tummel shows a flora almost identical with that already given as characteristic of the oak coppice."

R. Smith (*loc. cit.*) also stated that the ground vegetation of other examples of the Perthshire birch woods scarcely differs from the vegetation of the adjoining grassland or moörland. This, too, is the case with regard to some of the birch woods of the southern Pennines. Birches, when growing spontaneously, allow a great deal of light to pass through the canopy; and this fact, especially when coupled with the greater altitude of the birch woods, is sufficient to account for the smaller number of sciophytes or shade-loving plants and the greater number of photophytes or light-loving plants. The following lists, taken from two oak-birch woods of the Peak District, show the nature of the ground flora and vegetation:—

	Birch woods with moorland vegetation (Longdendale). Soil peaty and locally wet.	Birch woods with siliceous grassland vegetation (upper Derwentdale). Soil shaly and drier.
Sphagnum spp., and other mosses and liverworts	Abundant in the crevices between the boulders among which numerous fallen and rotting birch trunks litter the ground	Local
Blechnum Spicant	Occasional	Rare
Pteris aquilina	Rare to locally abundant	Rare to occasional
Salix caprea	Rare	Rare
S. cinerea	Rare to occasional	Rare to occasional
Betula pubescens	Very abundant	Very abundant
Quercus sessiliflora	Rare to locally abundant	Rare to occasional
Pyrus Aucuparia	Occasional	Rare
Crataegus Oxyacantha	Rare to abundant	Occasional
Galium saxatile	Occasional	Occasional
Calluna vulgaris, Vaccinium Myrtillus, V. Vitis-idaea	Completely covering many of the numerous rocks which strew the ground	Rare
Agrostis tenuis	Occasional	Occasional to abundant
Deschampsia flexuosa	Abundant	Rare to abundant
Molinia caerulea	Locally abundant	—
Festuca ovina	Occasional	Occasional
Nardus stricta	Occasional to abundant	Abundant
Carex pilulifera	Rare	Rare

Very few of the rare and characteristic herbaceous species of the Scottish birch woods occur in Great Britain so far south as Derbyshire. For example, the following species, which occur in the Scottish birch woods (R. Smith, 1900 *b*) are absent from Derbyshire:—

Pyrola secunda	Linnaea borealis
P. rotundifolia	Corallorrhiza trifida (=C. innata)
Moneses uniflora (=M. grandiflora)	Goodyera repens

Of the above, the coral-root stops at the Border: Goodyera almost stops at Cumberland, but has outlying stations in Yorkshire and Norfolk: *Moneses* (*Pyrola*) *uniflora* is unknown in England: *P. secunda* and Linnaea are rare in northern England; and *P. rotundifolia* is a very local plant throughout southern Britain. *Listera cordata* and *Trientalis europaea*, which are found in birch woods in Scotland, exist on the Pennines only as moorland plants[1]; and, even on the moors, they are rare and local. *Pyrola media* and *P. minor* appear, in fact, to be the only species of this class which are typical of both the Scottish birch woods and the upper woodland zone of northern England; and even these species are rare and local throughout the whole of England.

The Primitive Birch-Forest

Judging from the timber which is not infrequently found buried under the peat of the Pennines, it is certain that in former times a very extensive upland zone of birch woods existed on the Pennines; and the meagre birch woods which now occur on the Pennines are to be regarded as the vestiges of a former widespread plant association (cf. Smith and Rankin, 1903: 160). Although birches are quite commonly met with under the peat on certain of the moors of the district, one hesitates to refer to such a layer as a continuous forest bed. The layer is

[1] It has recently been stated (Williams, 1910: 127) that *Trientalis europaea* grows "in woods" at Halifax, which is the southern British limit for this plant. As a matter of fact, the plant in that locality is confined to a small space where it grows among bilberry and mat-grass on a treeless hill side. Wheldon and Wilson (1907: 239) state that on the Pennines further north the plant grows on "moorlands amongst bilberry, bracken, and heather."

not continuous under the peat of the southern Pennines, at all events: the birch remains do not exist, for example, on the exposed ridges, and they are absent from certain exposed hill sides: they frequently follow the hollows worn out by the streams; and, at their highest limit, they are practically limited to the stream banks. It seems to me unreasonable to elevate these discontinuous birch remains almost to the rank of a geological formation, as is done by some writers on peat; and it seems best to speak of them simply as the remains of a former birch forest, since their occurrence is exactly what one would expect them to be assuming they are the remains of a thin and open forest which once occurred at the upper local limit of woodland.

This ancient birch forest is wholly a post-glacial affair; and the reduction in altitude of the forest limit illustrates what is perhaps a general law that in any district where a forest exists at its extreme limits, climatic or otherwise, the forest will as time goes on exhibit retrogressive tendencies. The latter are usually intensified by human interferences, such as by felling and by the grazing of domestic animals, and, on the other hand, they may be retarded by human interference, as by the careful replanting of the indigenous trees; but, left to itself, any forest which exists at its climatic or edaphic limits will, in all probability, become degenerate in time. The causes of this degeneration are discussed rather more fully in the next chapter (see page 91).

It seems to be the case that, in this primitive Pennine birch forest, the Scots pine (*Pinus sylvestris*) occurred. However, as pine timber is only rarely met with under the peat of the Pennine moors, and as birch timber is abundant, it is impossible to postulate a general zone of pine forest at a different altitude from the birch forest. Probably the pine occurred much more rarely than the birch, either as an occasional associate in the birch association, or it formed smaller associations or societies here and there. On these assumptions, the ancient forest on the upper slopes of the Pennines would be regarded as part of the forest region of north-western Europe, but not, as is the case of the woodlands with birch and pine in southern England, as part of the forest region including the north German plain. The pine probably became extinct here at an early date; and the existing trees have, in all probability, either been planted, or they are the descendants of planted trees. The latter is much the more rare occurrence, as few pine seedlings are met with in this district.

ASH WOODS OF *FRAXINUS EXCELSIOR*

Ash woods are characteristic of the slopes of the limestone hills of the west and north of England. In previous British vegetation memoirs, woods of this type have been described on the limestone of the mid-Pennines (Smith and Rankin, 1903), in Westmorland (Lewis, 1904 *a*), and in Somerset (Moss, 1907 *a*). They do not appear to have been described by continental plant geographers. Smith and Rankin (1903: 168 *et seq.*) mentioned three sub-types. The first of these, which they termed "scar woods," occurs on the slopes of hills of the Carboniferous Limestone: such "scar woods" are more of the nature of scrub than of woodland. "The hazel [*Corylus Avellana*] is the dominant element most commonly found; but the ash [*Fraxinus excelsior*] occurs frequently, and sometimes close enough to reduce the hazel to a subdominant form....As a rule, the ash occurs but sparingly, because, being almost the only timber tree in the limestone dales, it is generally removed. In this district, few of the woods receive any attention; and little is done to check disforesting. Almost all the scar woods are therefore to be regarded as shrubby thickets" [*i.e.*, as scrub]. These ash and hazel scrubs are interesting in that they still furnish the habitat of that rare British orchid, the lady's slipper (*Cypripedium Calceolus*). The second sub-type mentioned by Smith and Rankin occurs at lower levels on the Permian or Magnesian Limestone and in the bottoms of the dales of the Carboniferous Limestone, and consists of ash woods which have been much altered by planting beech, oak, sycamore, pine, and larch. The third sub-type is termed a "hazel copse of the Permian," and is made up of scrub occurring on the Permian limestone. Lewis gives only a brief account of the woods of the district which he investigated; and it is not easy to relate them to general woodland types. However, the "birch woods" mentioned by him (1904 *a*: 319), judging from their occurrence at comparatively high altitudes on limestone and from the list of associated species, would appear to belong to the general ash type, and to the association of ash-birch woods (cf. p. 40). In Somerset (Moss, 1907: 41), ash woods are well developed on the slopes of hills of Carboniferous Limestone, of the Dolomitic Conglomerate, and of the Jurassic limestones. The "oak-hazel woods" of Somerset (Moss, 1907: 51) also are to be regarded as conforming to the general ash wood type (see Watson, 1909; and Moss, Rankin, and Tansley, 1910: 138). "Oak-hazel woods," *i.e.*, woods with oak standards and much hazel coppice, are of very general occurrence throughout southern England. Some of them have been derived from oak (*Quercus Robur*) woods, and others from ash-oak woods. The "oak-hazel woods" of Somerset must be referred to the latter class, because they contain among the coppiced layer a great deal of ash, which would spring up as standard trees if not coppiced, and because their "ground flora resembles the more shady portions of the ash wood" (Moss 1907: 52).

In the Peak District, typical ash woods occur on the slopes of the hills of the Carboniferous or Mountain Limestone (see figures 9, 10 and 17). None is represented in the northern area, as there no limestone rocks occur. In the southern area, they are well represented, especially in Wye dale and Lathkill dale.

The Carboniferous Limestone rocks of north Derbyshire form a plateau which attains a height of about 1550 feet (472 m.), and whose average height is perhaps 1200 feet (366 m.). The plateau is dissected by numerous valleys or "dales," most of which are streamless. The limestone dales of the Pennines are comparable with the gorges and coombes of the Mendip Hills of Somerset, both from the standpoint of the geology and that of the vegetation. The dales of Derbyshire descend from the plateau, and the ash woods begin to appear on the slopes at an altitude of about 1000 feet (305 m.), above which altitude scrub occurs, but no genuine woods. The woods continue to the bottoms of the dales, which here descend to about 250 feet (76 m.). This is much lower than any of the ash woods or scrub on the Carboniferous Limestones of the mid-Pennines; and, as in Somerset, the lower altitude permits of a better development of the dominant tree and the more characteristic shrubs and ground species of the ash woods.

It is probable that at some past time, the whole of the

Copyright *W. B. Crump*

Figure 9.

Ash Wood of *Fraxinus excelsior.*

General view. The wood clothes steep, rocky slopes of Carboniferous Limestone. In the foreground is a pool with Water-crowfoot and with marginal marsh plants.

limestone slopes and the more sheltered portions of the limestone plateau were covered by a primaeval ash forest, just as similar places on the sandstones and shales were once covered by forests of oak (*Quercus sessiliflora*) and birch (*Betula pubescens*). The numerous place-names including the word "ash" indicate that the dominance of the ash in the Peak District is of long standing. Of such names, one may mention Ashwood dale, Ashford dale, Money Ash (= many ash), and, on the edge of the plateau at the woodland limit, One Ash.

On the Chalk rocks of the south and east of England, the ash is a very abundant and characteristic plant, though its dominance in woods is apparently confined to their south-western margin, where ash woods occasionally occur (cf. Moss, Rankin, and Tansley, 1910: 137).

The recognition of the ash woods in England may fairly be claimed as a result of the method of vegetation survey, as their occurrence had apparently been quite overlooked both by foresters and botanists; and, as already stated, ash woods are undescribed for the continent of Europe. As Elwes (1908, iv: 870) has stated that the ash is probably the only hardwood which, at the present time, it pays to cultivate, it is obviously a matter of economic importance to note the distribution of spontaneous ash woods.

Although many of the ash woods have been interfered with, there can be no doubt that they represent the typical and natural vegetation of the calcareous hill slopes of northern and western England. Some of the ash woods show no signs of planting, and possess, in fact, all the attributes of a primitive plant association. The ash produces ripe seeds; and seedlings in all stages occur in abundance. The land agents and keepers of the ash woods assert that the ash is not planted, but that it springs up everywhere "like a weed." Many of the slopes on which the ash woods occur are too rocky and precipitous to have ever been enclosed as farmland (figure 9); and even on the less rocky slopes where the woods have degenerated into scrub and grassland, the land is not always reclaimed, but often remains uncultivated. Further, the associated trees, shrubs, and ground species are such as botanists agree in regarding as members of the primitive flora of the country. It is legitimate and reasonable, therefore, to regard the ash woods as primitive.

The ash woods on the limestone slopes have the same altitudinal range as the oak (*Quercus sessiliflora*) woods on the slopes of the sandstones and shales; that is, they range from the valley bottoms up to about 1000 feet (305 m.).

Semi-natural Woods and Plantations on the Limestone Slopes

Some of the slopes of the limestone hills with a deeper and a damper soil are utilized by the foresters for the growth of marketable timber; and the beech (**Fagus sylvatica*), the sycamore (**Acer Pseudoplatanus*), the larch (**Larix decidua*), and other trees are planted. Owing to the dip of the rock-strata, one side of narrow valleys is usually damper than the other; and hence it is unusual to find that the opposite sides of the dales have quite similar vegetation. One may find, for example, that the damper slope is planted up with beeches, sycamores, larches, and other introduced trees, and that the opposite side is characterized by a perfectly spontaneous ash wood; and thus it is in parts of Lathkilldale. In some cases, where alien trees have been planted on the site of a previous ash wood, the primitive flora lingers on for some time. For example, the lily-of-the-valley (*Convallaria majalis*) and the broad-leaved helleborine (*Helleborine latifolia*) still linger on, but do not flower freely, under introduced beeches in Lathkilldale. In other cases, the alien trees have been planted on calcareous grassland. In such plantations, one does not find the rarer and more characteristic plants of the ash woods; but some of the more general and ubiquitous woodland species, such as *Geum urbanum* and *Lychnis dioica*, sooner or later invade them. On the accompanying vegetation maps, the great abundance of introduced trees is, where possible, indicated by the initial letter of the alien tree being planted over the woodland colour; and thus it is often possible to infer from the maps whether alien trees have been planted in a wood or not. The maps, therefore, have a far greater value to foresters than any previously constructed maps, such as the Ordnance maps or the small scale maps issued by various publishers, as even the best of these maps do not attempt to distinguish more than deciduous woods and coniferous woods; and even this simple distinction is sometimes made inaccurately. Further, none of these maps attempts to distinguish between natural and semi-natural woods on the one hand and obviously artificial plantations on the other.

Trees and Shrubs

The ash (*Fraxinus excelsior*) is dominant throughout the length and breadth of the ash woods (see figure 9); and in them it is not confined, as it is in the oak woods, to the damper situations. It seems clear that, in any given natural station, the abundance of the ash is due to one of two causes, either to a high water-content or to a high lime-content. Some of the local foresters are of opinion that the timber of the ash grown on the limestone soils is harder and more durable than that grown on the wet, non-calcareous soils.

The two most frequent arboreal associates of the ash are the wych elm (*Ulmus glabra = U. montana*) and the hawthorn (*Crataegus Oxyacantha*), both of which are here more generally distributed than in the oak or birch woods. The elm is more abundant at the lower altitudes and in the damper situations (see figure 10), the hawthorn in the drier situations and at the higher altitudes. When the ash, the most valuable timber tree of the dales, is removed or dies out in a degenerating wood, the elm or the hawthorn, as the case may be, becomes locally subdominant; and societies of elm and hawthorn are as characteristic of the ash woods as birch and alder societies are of oak woods. On the vegetation maps, these societies are indicated by the same colour as the ash association of which they form a part; but, where practicable, the initial letter or letters of the genus of the locally subdominant tree is printed on the general woodland colour. An example of a society of wych elms occurs in upper Middleton Dale; and hawthorn societies are typical of most of the upper parts of drier dales.

Two conifers are native in the ash woods. One of these, the juniper ("*Juniperus communis*") is very rare, and apparently confined to one place: the other, the yew (*Taxus baccata*) is not common; but small specimens occur here and there on the ledges of limestone cliffs in the ash woods. It is rather curious that these plants should be so uncommon here, as they are very much more abundant further south, as on the Chalk of south-eastern England, and on limestone further north, as in north-west Lancashire.

Of introduced conifers, the larch (**Larix decidua*) and the Scots pine (**Pinus sylvestris*) are locally very abundant; but no evidence has been obtained that either of these trees rejuvenates itself from self-sown seed.

The aspen (*Populus tremula*) is the only indigenous poplar of the Peak District. It is decidedly uncommon on the whole; but occasionally, as in Cressbrookdale, aspen societies occur.

Of willows which are certainly indigenous, there are the crack willow (*Salix fragilis*), the osier willow (*S. viminalis*), and the sallows (*S. caprea* and *S. cinerea*); but *S. alba, S. triandra, S. pentandra, S. purpurea,* × *S. Smithiana* also occur by some of the stream sides. *S. aurita* and *S. repens* appear to be absent from the limestones.

The hazel (*Corylus Avellana*) is a very abundant and characteristic shrub, more so even than in the oak woods. Dense thickets of hazel frequently occur, especially in the subordinate scrub associations (see next chapter).

The alder (*Alnus glutinosa*) is even less abundant and less characteristic in the ash than in the oak woods; but locally it forms societies at the bottom of some of the damper dales, as in Cressbrookdale.

Birches are as rare as oaks in the ash woods of the Peak District, and are perhaps not indigenous. **Betula alba* and **B. pubescens* have both been planted in Haydale (Monsaldale), along with beeches and conifers, on the site of a former ash wood. The absence of oaks and birches from the ash woods of this district is interesting; as, in other parts of England, both trees occur more or less abundantly in ash woods.

Oaks are very rare and perhaps not indigenous on the Carboniferous Limestone of the Peak District. In the Wye valley, which is locally well wooded, only about half a dozen oaks were noted; and these did not occur in the more primitive of the ash woods, but only among trees which were obviously introduced, as in parklands and plantations.

The beech (**Fagus sylvatica*) is planted abundantly, but does not appear to be indigenous on the Pennines.

Several species of Ribes (*R. Grossularia, R. alpinum,*

Copyright *W. B. Crump*

Figure 10.

Ash Wood of *Fraxinus excelsior*.

A society of Wych Elms (*Ulmus glabra* = *U. montana*). The ground vegetation to the left of the footpath consists very largely of Dog's Mercury (*Mercurialis perennis*).

R. nigrum, R. rubrum) also occur; but these, with the possible exception of *R. alpinum*, are perhaps not indigenous.

Ericaceous undershrubs are totally absent from the ash woods; and this appears to apply to all the woods of the ash and beech associations (see page 40) throughout the country.

One of the most noticeable features of ash woods, both here and elsewhere, is the large number of arboreal and shrubby species which occur in the association. The shrubs are sometimes very dense, and almost impenetrable. The following species, which are absent or nearly absent from the oak and birch woods, are characteristic of the ash woods of the Peak District:—

"Juniperus communis" (rare)	Euonymus europaeus
Taxus baccata (local)	Rhamnus catharticus
Populus tremula (local)	Tilia cordata (rare)
Ribes alpinum (local)	Daphne Mezereum (rare)
Pyrus Aria (rare)	D. Laureola (local)
Rosa spinosissima (local)	Cornus sanguinea
R. micrantha (rare)	Ligustrum vulgare

Herbaceous Vegetation

The two most characteristic ground societies of the oak woods, namely the hair-grass society and the soft-grass society (see pages 55 and 56), do not occur at all in the ash woods. The hair-grass society is, it will be remembered, characteristic of those portions of the oak and birch woods whose soils have a high content of acidic humus; and the soft-grass society occurs in the drier parts of the oak woods whose soils have a lower, but still a decidedly appreciable proportion of acidic humus. Such humus does not accumulate in the ash woods.

The ash woods cannot be separated from the oak woods on the basis of differences in the water-content of the soil of the two plant communities; for in each case there is a range from very wet to very dry soils. The lime-content in the two cases, however, is always strikingly different; and there are no soils in the ash woods with a high content of acidic humus such as very frequently characterise the soils of the oak and birch woods. The following divisions of the ground vegetation will illustrate the range in habitat within the ash woods of the district.

1. Marshy places. In marshy places, which occur in the ash woods by stream sides, at the bottoms of some of the streamless dales, and in places where springs arise, such moisture-loving plants as the following occur, in addition to such indigenous trees as the ash, the alder, and the crack willow:—

Trollius europaeus (local)
Caltha palustris
Spiraea Ulmaria
Geum rivale
G. rivale × urbanum
Epilobium hirsutum
Myosotis palustris
Mentha aquatica
Valeriana officinalis
Petasites ovatus
Cnicus heterophyllus
Phragmites communis
Phalaris arundinacea
Scirpus compressus (rare)
Sparganium ramosum
Orchis maculata

2. Damp places. Other parts of the ash woods although not really marshy, are nearly always very moist; and such places, like similar ones in the oak woods, have a rich and varied ground flora. The trees are here usually well grown; and the wych elm is frequently abundant (see figure 10). Sheets of wood-garlic (*Allium ursinum*) and of the lesser celandine (*Ranunculus Ficaria*) are characteristic. The following is a selected list of the ground species of such parts of the ash woods:—

Nephrodium Filix-mas
Lychnis dioica
Anemone nemorosa
Ranunculus Ficaria
Trollius europaeus (local)
Aquilegia vulgaris (local)
Fragaria vesca
Geum rivale
G. rivale × urbanum
Oxalis Acetosella
Polemonium coeruleum (local)
Myosotis sylvatica
Lamium Galeobdolon
Asperula odorata
Valeriana officinalis
V. dioica (local)
Campanula latifolia (local)
Cnicus palustris
C. heterophyllus (local)
Deschampsia caespitosa
Bromus ramosus
Triticum caninum (local)
Hordeum europaeum (=H. sylvaticum) (local)
Carex sylvatica
Arum maculatum
Allium ursinum
Orchis maculata
Habenaria virescens (=H. chloroleuca) (local)

3. Dry places. On soils which are drier than the preceding, and which, during the summer months, may in fact become temporarily very dry, expanses of dog's mercury (*Mercurialis perennis*) often occur; and this plant is here frequently associated with the tiny moschatel (*Adoxa Moschatellina*). At

the beginning of April, in the Derbyshire dales, the dog's mercury is about three inches high: its leaves are beginning to unfold; and a few stamens are ripe. At this time of the year, the moschatel is here flowering abundantly, and is almost hidden by the young shoots of the dog's mercury. In the fairly dry portions of the ash woods of the Peak District, this ground society of dog's mercury and moschatel is a characteristic feature. The society is an excellent example of what Woodhead (1906: 345) would term a "complementary" society, as the roots of the dog's mercury reach down to lower layers of soil than the roots of the moschatel, whilst the small and delicate shoots of the Adoxa receive their necessary shade from the larger and more vigorous shoots of Mercurialis. Before the end of June, Adoxa has entered on its long period of dormancy; and the dull green leaves of the dog's mercury, hiding its ripening berries, occur in extensive and monotonous stretches. It may, therefore, be said that the roots of the two species are edaphically complementary and the shoots seasonably complementary. In the oak and birch woods, the dog's mercury occurs in more or less local patches, and Adoxa is extremely rare; whilst the Mercurialis-Adoxa society does not occur.

The dog's mercury is much more abundant, especially as a social species, in English woods on calcareous soils than in those on non-calcareous soils; and this is a partial confirmation of an observation made by Thurmann (1849) who mentions the plant as one of fifty "xerophilous" plants typical of "dysgeogenous" or calcareous soils.

Still drier parts of the ash woods are characterized by stretches of ground ivy (*Nepeta hederacea*) which remains green throughout the whole year and which flowers from early spring to late summer. If the ground is stony and composed of old screes, taller herbs occur, such as the hairy St John's wort (*Hypericum hirsutum*), the nettle (*Urtica dioica*), and the wood sage (*Teucrium Scorodonia*). These plants form close herbaceous thickets in summer; and their dead stalks remain upright and rigid throughout the succeeding winter and spring. Locally, the lily-of-the-valley (*Convallaria majalis*) and the stone-bramble (*Rubus saxatilis*) form fairly extensive plant societies; and in these, the nodding melic-grass (*Melica nutans*) and *Helleborine atro-rubens* sometimes occur.

4. Rocky knolls. The very driest parts of the ash woods occur on the rocky knolls. Here the soil is extremely shallow; and in places the bare rock protrudes. Trees and shrubs are absent; and the absence of shade allows of the growth of saxicolous lichens and bryophytes, of such ephemeral species as *Arenaria serpyllifolia*, *Erophila verna*, and *Saxifraga tridactylites*, and of such dwarf perennials as *Sedum acre* and *Thymus Serpyllum*. Such a community does not, except in a topographical sense, belong to a woodland association at all, and is to be regarded as an outlier of another association.

Limestone screes and cliffs also occur in the midst of the ash woods. These, if damp, become in time clothed with the vegetation of the ash woods; and, by comparing several such localities, it is possible to gain some idea of a progressive succession from bare screes and cliffs to a closed ash association. Such a succession supplies the reason, a historical one, why such plants as the mossy saxifrage (*Saxifraga hypnoides*) and the limestone polypody (*Phegopteris Robertiana*) are sometimes found on old screes in the midst of existing ash woods.

Comparison of the Woodland Plants of the Southern Pennines

The numerous trees and shrubs which occur in the ash woods and which are absent from the oak and birch woods have already been mentioned (see page 71). To the species of this class there mentioned, the following herbaceous plants may be added:—

Polypodium vulgare
Phegopteris Robertiana
Phyllitis Scolopendrium
(=S. vulgare)
Asplenium Trichomones
A. Adiantum nigrum
A. Ruta-muraria
Cystopteris fragilis
Helleborus viridis
"H. foetidus"
Cardamine impatiens
Draba muralis
Sedum Telephium
Saxifraga hypnoides
S. Tridactylites
Poterium Sanguisorba
Geranium lucidum
G. sanguineum
Hypericum montanum
H. hirsutum
Helianthemum Chamaecistus
Viola hirta
V. sylvestris
Pimpinella major
Polemonium coeruleum
"Lithospermum officinale"
Satureia Acinos
S. vulgaris
(=Calamintha Clinopodium)
Origanum vulgare

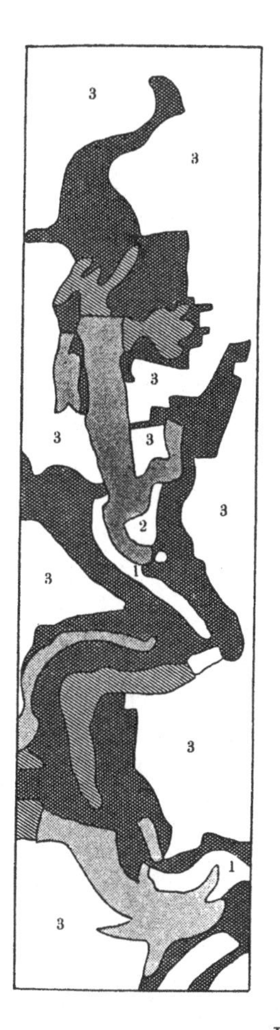

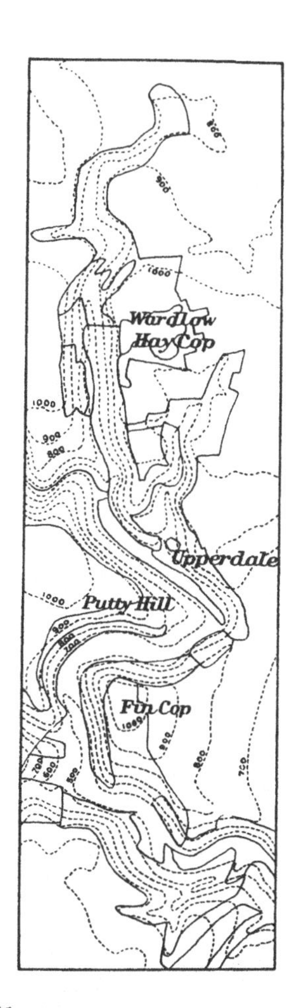

Figure 11.

Maps of Cressbrook Dale.

Left-hand map. The unshaded parts ( ) consist of permanent pasture :—(1) of valley bottom alluvium, (2) of intakes from the hill-slopes, (3) of the plateau. The cross-hatched parts ( ) consist of calcareous grassland. The hatched parts ( ) consist of calcareous scrub. The stippled parts ( ) consist of ash woods. *Right-hand map.* The same area with contour-lines in feet (1000 feet = 302 m.). The rocks consist entirely of Carboniferous Limestone.

"Atropa Belladonna"	Centaurea Scabiosa
Galium verum	Picris hieracioides
G. sylvestre	Hieracium spp.
"Rubia peregrina"	H. sciaphilum
Dipsacus pilosus	H. britannicum
Scabiosa Columbaria	Carex ornithopoda
Campanula Trachelium	Scirpus compressus
Eupatorium cannabinum	"Polygonatum multiflorum"
Inula squarrosa (=I. Conyza)	"P. officinale"
Arctium nemorosum	Helleborine atro-rubens (=Epipactis atro-rubens)
Serratula tinctoria	"Orchis pyramidalis"

On the other hand, the following species are found in the oak and birch woods, and are absent from the ash woods of the Peak District:—

Trees and shrubs

Salix aurita	Quercus sessiliflora
Betula pubescens	Rhamnus Frangula

Undershrubs

Salix repens	Erica cinerea
Ulex Gallii	Calluna vulgaris
U. europaeus	Vaccinium Myrtillus
Cytisus scoparius	V. Vitis-idaea

Herbaceous species

Equisetum sylvaticum	Jasione montana
Cryptogamma crispa	Wahlenbergia hederacea
Blechnum spicant	Gnaphalium sylvaticum
Nephrodium montanum	Senecio sylvatica
N. spinulosum	Holcus mollis
Corydalis claviculata	Deschampsia flexuosa
Pyrola minor	Molinia caerulea
P. media	Carex binervis
Scutellaria minor	C. helodes (=C. laevigata)
Digitalis purpurea	Luzula sylvatica
Melampyrum pratense (agg.)	Orchis ericetorum
Galium saxatile	

Several species, whilst occurring both in the ash and in the oak and birch woods, ascend to higher altitudes in the former than in the latter; and the following are examples of such species:—

Ulmus glabra
(=U. montana)
Sisymbrium officinalis
Geum urbanum
Rubus caesius
Rosa arvensis
Vicia sepium
Acer campestre
Sanicula europaea
Caucalis Anthriscus
Chaerophyllum temulum
Cnicus heterophyllus
Phragmites communis
Arrhenatherum elatius
(=A. avenaceum)
Brachypodium gracile
Arum maculatum
Allium ursinum
Tamus communis
Orchis mascula

The ash woods are much richer in species than the oak and birch woods, in spite of the fact that the species characteristic of soils containing acidic humus are abundant in the latter woods and absent from the former woods. Indeed, the floristic wealth of the ash woods is even greater than mere lists indicate, for several of the species which are rare and local in the oak and birch woods are more abundant and general in the ash woods. The following species belong to the latter class:—

Populus tremula
Mercurialis perennis
Trollius europaeus
Aquilegia vulgaris[1]
Sisymbrium officinalis
Rubus saxatilis[1]
Rosa tomentosa
R. glauca
R. arvensis
Pyrus Aria[1]
Geranium sylvaticum[1]
Polygala vulgaris
Tilia cordata[1]
Acer campestre
Primula vulgaris
Myosotis sylvatica[1]
Adoxa Moschatellina
Valeriana officinalis
Campanula latifolia[1]
Cnicus heterophyllus[1]
Poa nemoralis
Melica nutans[1]
Agropyrum caninum
Festuca sylvatica
Hordeum europaeum[1]
Carex sylvatica[1]
Paris quadrifolia[1]
Convallaria majalis[1]
Helleborine latifolia
(=Epipactis latifolia)
Listera ovata
Orchis mascula
Habenaria virescens
(=H. chlorantha)

Some species which are characteristic of the ash woods of Yorkshire (cf. Smith and Rankin, 1903) do not occur so far south as Derbyshire, and apparently find the intervening non-calcareous soils an effectual barrier. *Actaea spicata,*

[1] These species are not recorded by Linton (1903) for any of the oak woods of Derbyshire; but they occur in such woods on the eastern slopes of the Pennines a little to the north of the Peak District.

Melampyrum sylvaticum, and *Cypripedium Calceolus* are examples, though, judging from an old record, cited in Linton's flora (1903 : 274), the last-named species seems to have occurred formerly in Derbyshire.

The following plants are found in ash woods of Derbyshire, but do not occur so far south as Somerset, and are hence absent from woods of the ash type in the latter locality:—

Stellaria nemorum	Campanula latifolia
Cardamine amara	Cnicus heterophyllus
Trollius europaeus	Melica nutans
Geranium sylvaticum	Festuca sylvatica
Polemonium coeruleum	Hordeum sylvaticum
Myosotis sylvatica	Carex ornithopoda

On the other hand, the following plants occur in woods of the ash type in Somerset, but have not been noticed during the present survey in those of the hills of the Peak District:—

Clematis Vitalba	Viburnum Lantana
Aconitum Napellus	Calamagrostis Epigejos
Euphorbia pilosa	Colchicum autumnale
E. amygdaloides	Cephalanthera grandiflora
Lithospermum purpureo-coeruleum	Ornithogalum pyrenaicum

The autumn saffron (*Colchicum autumnale*) is indigenous in pastures on the Permian limestone; but it is not a woodland plant in the north of England, as it is in Somerset and Cambridgeshire.

The following lime-loving and shade-loving species occur on the lowland Permian limestone tract to the east of the Pennines, but are absent from the woods of the Carboniferous Limestone of the Peak District:—

Astragalus glycyphyllos	Viburnum Lantana
Galium Mollugo	Calamagrostis Epigejos

Generally, it is clear that the ash woods occurring on the calcareous soils of England are richer in species than the oak and birch woods occurring on the non-calcareous soils, and that of species common to both types of wood, many are more abundant and ascend to higher altitudes there than in the oak and birch woods. These facts can scarcely be held to support a statement sometimes made that calcium carbonate acts deleteriously on plants. Woodhead (1906: 396) states that his study of the woods around Huddersfield "indicates that in this district, the physical properties of the soil and accompanying conditions play a more important part in determining the character of the plant associations and the distribution of species than the chemical composition of the soil." It has, however, to be remembered that the district investigated by Woodhead is quite destitute of calcareous soils. My own observations lead me to believe that in western Europe the presence or comparative absence of calcareous compounds in the soil is, directly or indirectly, a factor of prime importance in the distribution both of plant communities and of species, that within the limits of any particular district possessing only calcareous soils the water-content of the soil is a factor of great importance, and that within the limits of any particular non-calcareous area two soil factors have to be considered, namely, the amount of water and the amount of acidic humus. This view is essentially in harmony with that taken up by Graebner (1895, 1901, 1909, 1910), and by Nilsson (1902). Graebner has maintained that natural divisions of vegetation will only be reached by basing the classification on the richness or poverty of the soil, whilst the water-content of the soil furnishes a useful factor for the subdivision of the vegetation-divisions thus obtained. It is probable, however, in districts such as the higher Alps and in tropical countries, that other master-factors come into play having a more important effect on vegetation than the presence or absence of lime in the soil; and probably the same occurs in some aquatic habitats, such as rapidly flowing streams.

The following is a list of the constituent plants of the ash (*Fraxinus excelsior*) woods and of the oak (*Quercus sessiliflora*) woods of the Peak District. The frequency of each species is indicated by letters in the two columns:—dominant (d), subdominant (s), abundant (a), occasional (o), rare (r), very rare (vr), local (l), occasional to abundant (o to a), etc. Herbaceous species which are confined or almost confined to the more shady parts of the woods are printed in thick type, and those preferring marshy places are printed in italics. Species which are not indigenous but which have been planted either in the woods or on the sites of former woods are preceded by an asterisk.

	Woods of *Fraxinus excelsior*	Woods of *Quercus sessiliflora*
Trees, shrubs, and undershrubs		
"Juniperus communis"	vr	—
Taxus baccata	lo	—
*Pinus sylvestris	l	ld
*P. austriaca	—	l
*Larix decidua	ld	ld
*Abies excelsa	l	l
Populus tremula	r, la	vr
*P. canadensis	l	l
*P. candicans	—	l
"Salix pentandra"	vr	vr
"S. triandra"	vr	vr
S. fragilis	lr	lo
× S. Smithiana	lr	lr
*"S. alba"	vr	vr
*S. purpurea	r	vr
S. viminalis	l	l
S. caprea	o to a	o to a
× cinerea	l	l
S. aurita	—	la
× caprea	—	l
× cinerea	—	l
S. cinerea	r to o	o to a
var. aquatica	?	l
var. oleifolia	?	l
Corylus Avellana	a, ls	o, la
Betula pubescens	*l	o, la
forma denudata	—	la
var. parvifolia	—	r
*B. alba (= B. verrucosa)	*l	r
Alnus glutinosa	l	o, la
*Fagus sylvatica	ld	ld
*Castanea sativa	—	r
*Quercus Robur (= Q. pedunculata)	vr	l
*Q. Robur × sessiliflora	—	r
Q. sessiliflora	—	d
Ulmus glabra (= U. montana)	a	lo
*Ribes Grossularia	r	vr
R. alpinum	r	—
*R. nigrum	vr	—
*R. rubrum	vr	—
Pyrus Malus	r to o	r
P. Aucuparia	r to o	o, la
P. Aria	r	*vr
Crataegus Oxyacantha	a	o, la
Rubus Idaeus	r	r, la

	Woods of *Fraxinus excelsior*	Woods of *Quercus sessiliflora*
Trees, shrubs, and undershrubs		
R. fruticosus (agg.)	a	va
R. fissus	vr	r
R. Rogersii	—	vr
R. carpinifolius	r	o
R. Lindleianus	r	a
R. pulcherrimus	r	o
R. Lindebergii	o	o
R. Selmeri	—	a
R. rusticanus	a	r
R. Sprengelii	—	la
R. leucostachys	o	o
R. Radula	o	o
R. podophyllus	—	o
R. dasyphyllus	a	a
R. corylifolius	o	r
R. caesius	la	lr
Rosa spinosissima	l	—
"R. mollis"	r	r
"R. mollis × spinosissima" (=R. villosa)	r	—
R. tomentosa	o	r
R. micrantha	lr	—
R. obtusifolia	vr	vr
R. canina	o to a	o, la
var. lutetiana	o	r
var. senticosa	r	—
var. dumalis	a	o
var. vinacea	r	—
var. urbica	o	r
var. dumetorum	—	r
var. arvatica	r	r
var. caesia	r	o
R. glauca	o	r
R. arvensis	a	o, la
Prunus spinosa	o to a	r to o, la
P. insititia	vr	*vr
P. Padus	o to a	r to a
*P. avium	?	lr
P. Cerasus	?	?
Ulex europaeus	—	r to o
U. Gallii	—	l
Ilex Aquifolium	lr	r to a
Euonymus europaeus	r to o	—
Acer campestre	o to a	l
*A. Pseudoplatanus	la	la
Rhamnus Frangula	—	vr
R. catharticus	r to o	—
Tilia cordata	vr	vr
"Daphne Mezereum"	vr	—

	Woods of *Fraxinus excelsior*	Woods of *Quercus sessiliflora*
Trees, shrubs, and undershrubs		
"D. Laureola"	vr	—
Hedera Helix	a	o to a
Cornus sanguinea	r, la	—
Erica cinerea	—	r
Calluna vulgaris	—	la
Vaccinium Myrtillus	—	la
V. Vitis-idaea	—	r, la
Ligustrum vulgare	la	—
Fraxinus excelsior	d	r to la
Sambucus nigra	o to la	o to la
Viburnum Opulus	o	o
Lonicera Periclymenum	o	o
Herbaceous species		
Equisetum sylvaticum	—	*la*
var. *capillare*	—	l
Hymenophyllum peltatum	—	*vr*
Cryptogamma crispa	—	vr
Polypodium vulgare	vr	—
Phegopteris polypodioides	**vr**	**l**
P. Dryopteris	*r*	*r*
P. Robertiana	la	—
Nephrodium montanum	—	la
N. Filix-mas	o	o, la
var. paleaceum	—	la
N. spinulosum	—	l
N. aristatum (=N. dilatatum)	r	o, la
Aspidium angulare	**vr**	**vr**
A. aculeatum	**vr**	**vr**
Cystopteris fragilis	l	—
Athyrium Filix-foemina	**r**	**o, la**
var. **convexum**		**l**
Asplenium Trichomones	o, la	—
A. Ruta-muraria	o	—
Phyllitis Scolopendrium (=S. vulgare)	vr	—
Blechnum spicant	—	o
Pteris aquilina	lr	r to s
Urtica dioica	ls	la
Rumex nemorosus	**o**	**o**
R. Acetosa	*o*	*o*
R. Acetosella	r	la
Polygonum Bistorta	l	l
*Claytonia sibirica	—	l
Lychnis dioica	**a**	**r to a**
Stellaria nemorum	*r*	*r*
S. Holostea	**o**	**o**

	Woods of *Fraxinus excelsior*	Woods of *Quercus sessiliflora*
Herbaceous species		
Arenaria trinervia	**o**	**o**
Caltha palustris	*l*	*l*
Trollius europaeus	*l*	*lr*
Helleborus viridis	r	—
"H. foetidus"	r	—
Aquilegia vulgaris	**o**	**lr**
Anemone nemorosa	**a**	**la**
Ranunculus Ficaria	**o to s**	**la**
R. auricomus	**o**	**r**
Corydalis claviculata	—	r to o
*Iberis amara	l	—
Sisymbrium Alliaria	r to o	r
Cardamine amara	*r*	*r*
C. impatiens	r	—
C. flexuosa	**o**	**o**
C. pratensis	*l*	*l*
Draba muralis	l	—
*Arabis albida	l	—
Sedum Telephium	**r**	—
*"Saxifraga umbrosa"	l	—
S. hynoides	la	—
S. granulata	r to o	r
Chrysosplenium alternifolium	*lo*	*r*
C. oppositifolium	*la*	*la*
Spiraea Ulmaria	*l*	*l*
Rubus saxatilis	ls	vr
Fragaria vesca	**a**	**r to o**
Potentilla sterilis	o	o
P. erecta	r	o
Geum urbanum	o	lo
G. rivale	*la*	*la*
G. rivale × urbanum	*r*	*vr*
Poterium Sanguisorba	la	—
Vicia sepium	**o to a**	**r to o**
V. sylvatica	**r**	**vr**
Lathyrus pratensis	o	o
L. montanus	r	o
Geranium sylvaticum	**r**	**vr**
G. lucidum	la	—
G. Robertianum	o	o
G. sanguineum	r, la	—
Oxalis Acetosella	**o**	**o**
"Euphorbia amygdaloides"	**vr**	—
Mercurialis perennis	**o to s**	**o**
*Impatiens parviflora	—	l
*Hypericum calycinum	l	—
H. Androsaemum	vr	vr
H. perforatum	l	lr
H. hirsutum	o, la	—

	Woods of *Fraxinus excelsior*	Woods of *Quercus sessiliflora*
Herbaceous species		
H. montanum	r	—
H. pulchrum	r	o
Helianthemum Chamaecistus	la	—
Viola palustris		*l*
V. hirta	**o to a**	—
V. Riviniana	a	?
var. villosa	r	—
var. diversa	?	a
V. sylvestris	**r to o**	—
Epilobium hirsutum	*la*	*l*
E. parviflorum	r to o	r
E. montanum	o	o
E. angustifolium	l	l
Circaea lutetiana	**o**	**o**
var. **intermedia**	**r**	**r**
Sanicula europaea	**o**	**r to o**
Ægopodium Podagraria	l	l
Pimpinella major	r, la	—
P. Saxifraga	r to o	r to o
Conopodium majus	o	o
Myrrhis Odorata	la	la
Chaerophyllum temulum	o	r to o
Anthriscus sylvestris	l	l
Angelica sylvestris	*o*	*o*
Heracleum Sphondylium	o	o
Caucalis Anthriscus	la	la
Pyrola media	—	vr
P. minor	—	vr
"Monotropa Hypopytis"	**vr**	**vr**
Primula veris	r to o	vr
P. vulgaris	**o**	**r**
P. veris x vulgaris	**vr**	—
Lysimachia nemorum	*r*	*r to o*
Calystegia sepium	r	r
Polemonium caeruleum	*la*	—
"Lithospermum officinale"	r	—
Myosotis palustris	*la*	*la*
M. sylvatica	**o**	**r**
Myosotis arvensis	l	l
var. **umbrosa**	**o**	—
Ajuga reptans	**o**	**lo**
Teucrium Scorodonia	l	o
Scutellaria minor	—	*r*
Nepeta hederacea	la	r to o
Prunella vulgaris	o	o
Galeopsis Tetrahit	r to o	r to o
Lamium Galeobdolon	**o**	**r to o**
*L. maculatum	l	—

	Woods of *Fraxinus excelsior*	Woods of *Quercus sessiliflora*
Herbaceous species		
Stachys sylvatica	la	la
S. officinalis (=S. Betonica)	lo	lo
Calamintha Acinos	lr	—
C. vulgare (=C. Clinopodium)	**r to o**	—
Origanum vulgare	a	—
Thymus Serpyllum	la	—
Mentha aquatica	*l*	*l*
M. rotundifolia	*vr*	—
Solanum Dulcamara	*l*	*l*
"Atropa Belladonna"	vr	—
Scrophularia nodosa	*o*	*o*
"*S. alata*" (=*S. Ehrharti*)	*o*	*? r*
**Mimulus luteus*	*l*	—
Veronica montana	**lo**	**lo**
V. Chamaedrys	o	o
Digitalis purpurea	—	r to a
Melampyrum pratense	—	?
var. hians	—	r
var. montanum	—	r to o
Lathraea Squamaria	**r**	**r**
Galium Cruciata	la	r
G. Aparine	la	la
G. verum	la	—
G. saxatile	r	o to a
G. sylvestre	la	—
Asperula odorata	**la**	**la**
Adoxa Moschatellina	**o to a**	**vr**
Valeriana dioica	*r*	*r*
V. officinalis	*la*	
V. sambucifolia	*?*	*la*
Dipsacus pilosus	r	r
Campanula latifolia	**lo**	**vr**
C. Trachelium	lo	—
C. rotundifolia	l	r to o
Wahlenbergia hederacea	—	*r*
Jasione montana	—	r to o
Eupatorium cannabinum	*la*	
Solidago Virgaurea	r to o	r to a
Gnaphalium sylvaticum	—	r
Inula squarrosa	r	
Petasites ovatus	*la*	*lo to la*
Senecio sylvaticum	—	l
Arctium minus	r to o	r to o
A. nemorosum	**r to o**	—
Carduus palustris	*a*	*la*
C. crispus	r	vr

	Woods of *Fraxinus excelsior*	Woods of *Quercus sessiliflora*
Herbaceous species		
C. heterophyllus	*lo*	*vr*
Serratula tinctoria	o	—
Centaurea Scabiosa	o	—
Lapsana communis	l	l
Picris hieracioides	r	—
Crepis paludosa	*r*	*lo*
Taraxacum officinale (agg.)	o	o
Lactuca muralis	l	l
Hieracium spp.	la	r to o
H. boreale	?	r
H. britannicum	la	—
H. sciaphilum	la	
Anthoxanthum odoratum	o	o
Milium effusum	**r**	**r**
Agrostis tenuis	la	la
Holcus mollis	—	ls
H. lanatus	*o*	*o*
Deschampsia caespitosa	*la*	*la*
D. flexuosa	—	ls
Arrhenatherum elatius	la	r to o
Molinia caerulea	—	la
Melica nutans	**r to o**	**vr**
M. uniflora	**o**	**o**
Dactylis glomerata	lo	lo
Poa nemoralis	**r**	**vr**
P. trivialis	la	r to o
Festuca ovina	la	l
F. sylvatica	**vr**	**vr**
F. gigantea	**o**	**o**
Bromus ramosus	**o**	**o**
B. sterilis	r to o	r to o
Brachypodium sylvaticum	la	r
Agropyrum caninum	**o**	**r**
Hordeum europaeum (=H. sylvaticum)	**r to o**	**vr**
Carex remota	*r*	*lo*
C. pallescens	**vr**	**vr**
C. flacca	o	l
C. pendula	*r*	*r, la*
C. helodes (=C. laevigata)	—	*la*
C. sylvatica	**r**	**vr**
C. ornithopoda	la	—
C. pilulifera	vr	r to o
Scirpus sylvaticus	*vr*	*vr*
Arum maculatum	**o**	**r to o**
Juncus effusus	*l*	*la*
Luzula maxima	—	*la*

	Woods of *Fraxinus excelsior*	Woods of *Quercus sessiliflora*
Herbaceous species		
L. pilosa	o	o
Gagea lutea	*vr*	*vr*
Allium ursinum	ls	r to o
Scilla non-scripta	**r**	**o to s**
Polygonatum multiflorum	r	—
P. officinale	lr	—
Convallaria majalis	**la**	**vr**
Paris quadrifolia	**r**	**vr**
*Narcissus Pseudo-narcissus	l	l
*Galanthus nivalis	vr	vr
Tamus communis	**o**	**lo**
Iris Pseudacoris	*lr*	*lr*
Neottia Nidus-avis	**vr**	**vr**
Listera ovata	o	r
Helleborine latifolia	**o**	**lr**
var. **atro-viridis**	—	**lr**
H. atro-rubens	lr	—
Orchis mascula	**la**	**r**
"O. pyramidalis"	vr	—
Habenaria virescens (=H. chloroleuca)	**r to o**	**vr**
H. bifolia	vr	r
"Cypripedium Calceolus"	extinct	—
Musci		
Tetraphis pellucida	—	la
Catharinea undulata	a	o
Polytrichum formosum	—	la
Dicranella heteromalla	r	o
Dicranum scoparium	r	la
Campylopus flexuosus	—	l
Leucobryum glaucum	—	l
Fissidens bryoides	r	o
F. adantioides	o	r
F. taxifolius	o	r
Rhacomitrium lanuginosum	o	vr
Bryum capillare	o	o
Mnium hornum	—	o
M. undulatum	o	r
M. stellare	o	r
M. punctatum	r	o
Porotrichum alopecurum	o	o
Thuidium tamariscinum	o	r
Eurynchium praelongum	—	o
E. striatum	o	o
E. confertum	r	o
Plagiothecium denticulatum	a	a
P. Borrerianum	—	a
Hypnum Schreberi	o	o

	Woods of *Fraxinus excelsior*	Woods of *Quercus sessiliflora*
Hepaticae		
Hylocomium splendens	o	o
H. rugosum	r	
Fruillania Tamarisci	la	—
F. dilatata	la	—
Lejeunia serpyllifolia	la	—
L. calcarea	o	—
L. Rosettiana	o	—
Radula complanata	o	—
Porella platyphylla	o	—
Blepharozia ciliaris	—	o
Lepidozia reptans	—	a
L. setacea	—	o
Kantia arguta	—	r
K. Trichomonis	—	o
Cephalozia bicuspidata	—	a
Scapania nemorosa	—	o
Lophocolea bidentata	r	o
L. cuspidata	—	o
L. heterophylla	o	r
Plagiochila asplenoides	o	o
Jungermannia sphaerocarpa	—	o
J. ventricosa	—	o
Nardia scalaris	r	r
Fossombrosia pusilla	r	r
Pellia epiphylla	—	la
P. calycina	o	r
Aneura pinguis	r	r
Metzgeria furcata	r	vr
Fegatella conica	o	o
Lunularia vulgaris	r	r

CHAPTER III

SCRUB ASSOCIATIONS

Past and present upper altitudinal limit of trees. Buried timber in the peat. Degeneration of woodland. Distribution and character of the existing scrub. Scrub in other districts. Relation of the ground vegetation of woodland to retrogressive scrub. Progressive and retrogressive scrub. Comparison of the types of retrogressive scrub.

PAST AND PRESENT UPPER ALTITUDINAL LIMIT OF TREES

IT was pointed out in the last chapter that the upper altitudinal limit of oak and ash woods at the present time is in this district about 1000 feet (305 m.) and the upper limit of birch woods is about 1250 feet (381 m.). Isolated trees and patches of scrub, however, ascend to about 1550 feet (472 m.); and there can be no doubt that formerly trees ascended, in the Peak District, to about 1750 feet (533 m.) or 1800 feet (549 m.). These figures represent the upper limits of trees and woods on the highest hills alone: on the lower hills, the upper limits of trees and woods are not so high. For example, in the latitude of Halifax (a few miles to the north of the Peak District), the Pennines only rise to about 1550 feet (472 m.); and the present tree limit there occurs at about 1250 feet (381 m.) and the woodland limit at about 1000 feet (305 m.). According to W. G. Smith (1911: 20), in the Highland glens of Scotland, birch woods sometimes ascend to an altitude of 2000 feet (610 m.), in spite of the more northern latitude; but here mountains are massed together and rise to more than 3000 feet (915 m.). On Ben Nevis, the highest mountain (4400 feet = 1313 m.) in the British Isles, a tree is said to occur at 2700 feet (823 m.). Still further north, in Scandinavia, where the mountains are still higher, the trees commonly ascend to 3000 feet (915 m.).

It is clear, therefore, that the upper altitudinal limit of woods or of trees in any particular district cannot be accounted for merely by the factors connected with the altitude of that district. This point of view, although quite an old one, is frequently ignored.

Smith (1911: 13) says that "tree growth ceases wherever the wind attains such a force the formation of young shoots is prevented. The determination of this wind zone must always be considered in forestry. It cannot be defined as so many feet above sea-level. The action of the wind may be very marked on the coast itself....It is also the case in a hill-mass, that all the zonal limits of plants are lifted up." This is indeed quite true, and helps to explain the occurrence of woodland plants under the peat of Cross Fell (which rises to 2930 feet = 893 m.) at an altitude of 2400 feet (731 m.), whilst under the peat of the Peak of Derbyshire (which only rises to 2088 feet = 636 m.) woodland plants have not, during the course of the present investigation, been observed higher than 1800 feet (549 m.). It is not necessary to invoke post-glacial climatic changes to account for apparent discrepancies of distribution of this nature, for they can easily be paralleled by similarly apparent discrepancies in existing vegetation.

Warming (1909: 39) has stated that trees cease on mountains at the altitude where they break up into separate peaks. It seems highly probable that at the conclusion of the glacial period, this country was invaded by Arctic-Alpine species, and afterwards by forest or woodland species. Lewis (1905, etc.) finds two forest layers in some of the Scottish peats, so that perhaps there were two separate invasions of forest plants. However, only one such layer appears to be represented as a rule in the peat of the Peak District, though in other parts of England two or more layers of trees are found buried by peat. The primitive woods probably ascended the mountains up to or nearly up to the limit indicated by Warming. Perhaps there was above this primitive forest limit, a narrow girdle of climatic scrub and a still higher girdle of climatic grassland; but, as regards the present district, this is not certain. In any case, from that time to this, there has been a gradual lowering of the forest limit; and the scrub and grassland, which now characterise the higher slopes of the district (and indeed those of Europe generally) below actual Alpine limits, are largely the result of the retrogression or decay of the original forests.

Whilst this process of the lowering of the altitudinal forest limit in post-glacial times has, in my opinion, been essentially a natural process, it undoubtedly has in this, as in most localities, been greatly aided by the indiscriminate felling of trees by man and by the browsing of quadrupeds.

BURIED TIMBER IN THE PEAT

The most direct evidence regarding the former greater development of forest in the district is to be derived from a study of peat deposits. The number of memoirs dealing with this subject is legion; and practically all the writers have emphasized the view that the deposition of peat, in a large number of instances, has been preceded by the occurrence of forest. Equal emphasis, however, must be placed on two other facts. First, forests may degenerate and still no peat deposits may occur on the site of them; for example, the degeneration of woods on chalk rock or on limestone or on steep shaly slopes is not succeeded by peat deposits. Secondly, peat deposits may be laid down without the occurrence of any pre-existing forest; for example, peat is now filling up some of the Cheshire and Shropshire meres and the Norfolk broads; and in these localities it is clear that the vegetation preceding the deposition of the peat was a reed swamp with no arboreal plants; and also on the highest parts of the Pennine watersheds, peat may occur to the depth of twelve feet (363 cm.) or more without there being any trace of buried timber.

During the course of this vegetation survey, many instances of buried timber have been recorded on the field maps. The general inferences to be drawn from the facts are that tree trunks occur at the base of the peat of several of the heather moors and the lower cotton-grass moors, but that on the highest moors buried timber is frequently absent. Generally, it may be said that remains of trees are found under the peat in the more sheltered parts of the moor and are absent from the most exposed places. The buried trees which have been noted consist almost wholly of birch (probably *Betula pubescens*); but aspen (*Populus tremula*), oak (probably *Quercus sessiliflora*),

Copyright *W. B. Crump*

Figure 12.

A Sandstone Clough.

Siliceous grassland of Mat-grass (*Nardus stricta*) and silver Hair-grass (*Deschampsia flexuosa*) on the right. Scrub of Birches (*Betula pubescens*).

alder (*Alnus glutinosa*), hazel (*Corylus Avellana*), mountain ash (*Pyrus Aucuparia*), and willows (probably chiefly *S. cinerea*) occur more or less rarely; and in one locality Scots pine (*Pinus sylvestris*) was found.

The highest examples of buried timber consisted wholly of birch, and were encountered on the southern extremity of the plateau of the Peak at an altitude of nearly 1800 feet (549 m.); and generally it may be concluded that the buried timber proves that in former times trees ascended the southern Pennines about 200 feet (61 m.) or 250 feet (76 m.) higher than they do at the present time, that this ancient forest was composed principally of birches, and that more or less uncommon associates were the aspen, oak, the alder, the hazel, the mountain ash, the willow, and the Scots pine.

Degeneration of Woodland

There can be no doubt that a certain amount of the degeneration of the woodland of this district has been brought about by the indiscriminate felling of trees, the absence of any definite system of replanting, and the grazing of quadrupeds. It is doubtful, however, if these causes are quite sufficient to account for so great a lowering of the upper limit of forest as 250 feet (76 m.), and for so general a phenomenon. It must be remembered that the population of the remoter valleys, many of which are now treeless or almost so, is very small; and the district does not appear to have ever been a great grazing district.

The inability of certain forests to rejuvenate *per se* has been pointed out by many foresters and plant geographers. In discussing the causes of the succession of forest to heath in north Germany, Krause (1892) emphasized the view that the narrowing of the forest area has been largely due to errors in sylviculture, especially to the grazing of cattle in the forest. That such a factor is a *causa vera* in the degeneration of forests is indisputable. Graebner (1901), on the other hand, lays stress on the gradual impoverishment of the soil caused by the removal of the tree trunks, by the gradual washing out by rain of the soluble mineral salts originally present in the soil, and the spreading of heath vegetation on the forest floor consequent on the formation of moor-pan (*Ortstein*) in sour soils. Graebner pointed out that his explanation does not invalidate Krause's view as a partial explanation. Graebner's theory is a very reasonable one with regard to woods which occur on comparatively flat plains and plateaux; but it is scarcely satisfactory with regard to forests on many steep hill slopes, for, in such places, newer and richer soil from below is often exposed by denudation and occasional land slips may bring fresh soils from above. As the great majority of the degenerate woods of this locality are situated on such steep slopes, some additional explanation of forest degeneration must be sought.

Of course, it is well known that the seedlings of most trees fail to develop under dense shade; and, for this reason, some forests fail to rejuvenate. For example, in the High Engadine, in Switzerland, it has been established by means of long-continued observations that the forests of larch (*Larix decidua*) which partly cover the slopes and parts of the valleys of this part of the Alps do not everywhere regenerate themselves from seed. The seedlings of larch require abundant light; and this they do not always find beneath the old forest-growth. But the Arolla pine (*Pinus Cembra*) finds the conditions of light more favourable to its development. It sows itself abundantly and develops vigorously; so that under these special and rare conditions, the forest of Arolla pine will succeed the forest of larch without the intervention of man (cf. Flahault and Schröter, 1910; Rubel, 1911). However, no such explanation as this is applicable in the present district.

A matter which, in my judgment, is not as a rule sufficiently emphasized by plant geographers and foresters is that, in a closed plant community, seedlings, especially seedlings of plants with large seeds such as the oak and beech, are rarely found. On the other hand, open and (to a less extent) intermediate associations, if the general life-conditions are favourable, permit of invasion and rejuvenescence. For example, the elms near Cambridge produced an excessive quantity of fertile seeds in the summer of 1909. Many of these seeds germinated on more or less bare patches of soil, but not on the adjoining closed pasture-land. It follows that a wood whose carpet is fully occupied by closed ground societies does not tend to rejuvenate itself; and, as the more upland ash, oak, and birch woods of this district are, on the whole, characterized by such closed ground societies, it would seem that here is an additional reason which helps to explain the gradual degeneration of the forests of the Pennines. It is difficult, for example, to see how a close turf of silver hair grass (*Deschampsia flexuosa*) can be colonized by oak or beech seedlings; and, in fact, such seedlings are rarely seen in these situations. This fact is known to some foresters of the country; and use is made of their knowledge in that many of the woods and plantations of which they have charge have the ground kept more or less free of woodland "weeds."

The difficulty which larch seeds experience in germinating in closed herbage in the larch forests of the Altai Mountains has been pointed out by Krassnoff (1886) and quoted by Warming (1909: 316): "the herbaceous vegetation consists of species of Aconitum, Delphinium, Paeonia, Clematis, and others. Each year millions of larch seeds fall into this sea of herbage; yet only a few find places where they can germinate: the forest is apparently doomed to extinction."

The remarkable series of climatic changes within the historical period, which are invoked by certain writers to account for plant-successions, are always open to a certain amount of suspicion. In general, plant-successions, which have taken place since early post-glacial times and in a region of fairly uniform present-day climate, would seem to be explicable by changes in the physiographical and edaphic conditions of plant habitats.

Distribution and Character of the Existing Scrub

The existing woodlands, at their upper altitudinal limits, often pass imperceptibly into open scrub. On many of the hill-slopes of the remoter valleys, trees are more or less thinly scattered about; and it is, in fact, not always easy to decide whether or not a particular tract of vegetation should be considered scrub or poor woodland. Longdendale, Upper Derwent Dale, and Upper Cressbrook Dale furnish excellent examples of scrub. In some cases, the ground vegetation is grassy, in others heathy undershrubs are abundant. In some cases, the tallest plants are shrubs; and these sometimes form dense thickets: in others, shrubs are absent, and the uppermost layer is a very thin forest of more or less dwarfed and stunted trees. In nearly all cases, however, the scrub of this district appears to consist of retrogressive forest communities, and only rarely, as, for example, on certain fresh and newly formed soils beneath cliffs, of scrub progressing towards mature woodland. In the retrogressive scrub, a number of the more hardy ground species of woods still persist, such as the wood-rush (*Luzula pilosa*), the wood vetch (*Vicia sepium*), *Lathyrus montanum*, the wood violet (*Viola Riviniana*), the cow-wheat (*Melampyrum montanum*), and the ubiquitous bracken (*Pteris aquilina*); but their ultimate extinction, except perhaps in the case of the bracken, as the woodland or scrub vegetation degenerates still further towards grassland or heath or moor, appears certain.

Several of these areas still retain the place-name "wood," although now the name is most inappropriate; but as such areas occur within the primitive woodland zone, on more or less sheltered slopes often near the head of the cloughs (cf. figure 12) and dales (cf. figure 13), there need be no doubt that the place-name really indicates the former nature of the vegetation. It would appear to be true that, in districts which are capable on climatic and edaphic grounds of supporting woodland or true forest, the majority of the examples of open scrub are to be regarded as degenerate woods and as retrogressive associations. A study of numerous examples of such associations leads to the conclusion that the following successions have occurred and are still occurring in this district:

Succession I	Succession II	Succession III	Succession IV
Oak and birch woods on sandstone plateaux	Oak and birch woods on rocky sandstone slopes	Oak and birch woods on steep shaly slopes	Ash woods on limestone slopes
↓	↓	↓	↓
Scrub	Scrub	Scrub	Scrub
↓	↓	↓	↓
Siliceous grassland and heather	Siliceous grassland (with Molinia)	Siliceous grassland	Calcareous grassland
↓	↓		
Moors	Moors		

Copyright *W. B. Crump*

Figure 13.

A streamless Limestone Dale.

Calcareous grassland of sheep's Fescue-grass (*Festuca ovina*) in the foreground and on the right. Scrub of Hawthorns (*Crataegus Oxyacantha* = *C. monogyna*) on the left. The dale is streamless, and fenced with limestone walls.

Examples of Succession I occur on some of the Coal-measure plateaux on the eastern Pennines at an altitude of about 800 feet (244 m.), of Successions II and III in the cloughs of the sandstones and shales (cf. figure 12), and of Succession IV in the limestone dales (cf. figure 13).

The "scrub" of Crump (1904: xxxviii), the "clough thicket" of Smith and Moss (1903: 387), the "gill wood" and the "hazel copse" of Smith and Rankin (1903: 159 and 173), and the "ash copse" of Moss (1907 *a*: 44) are here included in the term scrub which is regarded as the English equivalent of the German "gebüsch."

Professor Diels (in Flahault and Schröter, 1910: 19) considers the use of vernacular names in plant geography very questionable. He maintains that such terms are ambiguous even in the language to which they belong, that to foreigners they are either meaningless or liable to misunderstanding, that even if such terms be once strictly defined they will become confused again, that they are permanently confusing to people unversed in phytogeography, that newly coined expressions (*e.g.*, "Hochmoor" and "high moor") are not truly indigenous terms and are most confusing to non-specialists, and that it is therefore desirable to have universal expressions in Latin or Greek, and to have these alone. With Diels' general position I have very much sympathy; but it is quite impossible, even if it be desirable, to abolish vernacular terms even when these do lead to some confusion. Diels specially singles out the English term "scrub" as a phytogeographical *nomen confusum*; and to this might be added the English terms "forest[1]," "heath[2]," and "swamp," and perhaps indeed every popular physiographical and phytogeographical term. It appears to me that the only course to adopt is to use vernacular names in the most frequently accepted sense, and, in addition, to use universal names which are not capable of misunderstanding.

[1] "Forest," in English, may signify almost any wild, open, uncultivated tract of land, not necessarily a tract of woodland, though historical documents prove that parts, at least, of the ancient British forests were tree-clad at some earlier period.

[2] Although, in English, a heath is usually a heather-clad tract of land, yet, in eastern England, the term is also used to denote a tract of calcareous pasture with no heather, as Newmarket Heath and Royston Heath; and in Somerset, it is used to designate tracts of deep and often wet peat.

SCRUB IN OTHER DISTRICTS

Clements (1905: 287) has maintained that "in forests, while many vegetation forms can still enter, none of these produces a reaction sufficient to place the trees at a disadvantage; and the ultimate forest stage, though it may change in composition, cannot be displaced by another." If my contention in the previous section of this chapter be sound, it follows that this generalization of Clements is not of universal application. In this district, and indeed in very many other districts, it would appear to be indubitable that woodland is frequently displaced by associations of scrub, grassland, heath, and moor. In all parts of the British Islands, there has, within the historical period, been a pronounced diminution of the forest area, a diminution which, in my judgment, is in addition to and apart from any artificial disforestation or any change of climate. The decay of forests in central Europe and the conversion of many of them into grassland and heaths is admitted by most phytogeographers; and there are not wanting authorities who have gone so far as to assert that prairies and even steppes have been derived from pre-existing forest (cf. Warming, 1909: 282), though it is difficult to accept this view, especially with regard to the origin of climatic steppes. In practically all cases of the ascertained conversion of forest into grassland, it would seem certain that an intermediate stage of open scrub occurred. It has also been urged by some plant geographers that some tropical forests have degenerated into savana-forest and scrub; and whilst this degeneration must obviously be accelerated by a diminishing rainfall, it is by no means improbable that the retrogressive succession may also take place in districts where such a decrease is imperceptible. In Great Britain, the conversion of woodland into scrub, and of scrub into grassland, heath, or moor is seen not only on the Pennines, but in Wales, in the Lake District, and in Scotland; and some of these districts have a mean annual rainfall of 80 inches (203 cm.) and occasionally more. Such successions are not exceptional in this country, but widespread and general; and whilst they are without doubt often due, in part, to artificial causes, it is at least conceivable that this is not always and wholly the case.

In districts where the rainfall is low, as on the borders of steppes and at very high altitudes, where the amount of precipitation is insufficient to permit of the growth of large trees, there can be no doubt that static, climatic scrub occurs; and, on certain very dry soils in moderately rainy localities, it is also certain that static, edaphic scrub occurs.

The relations of the most important types of scrub are shown in the following scheme:—

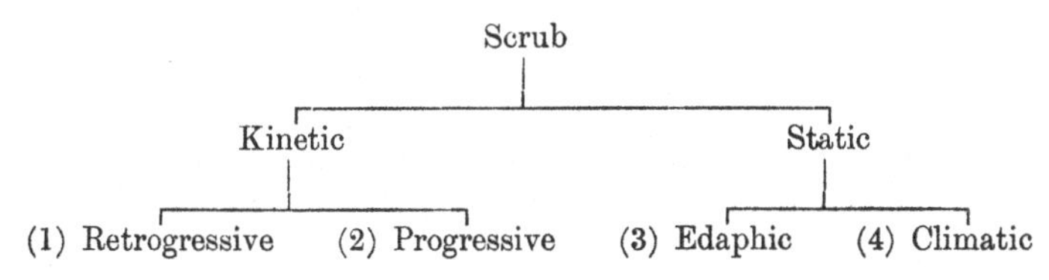

Of these, the examples of scrub met with in the Peak District belong mostly to the first type; and no doubt the great majority of the British examples of scrub should be placed in the same class.

Relation of the Ground Vegetation of Woodland to Retrogressive Scrub

Although nearly all the examples of scrub on the Pennines belong to the retrogressive type, they are important, and no account of the vegetation of a district which failed to account for them, could be regarded as complete. At the same time, it does not appear to be reasonable to regard subordinate associations such as retrogressive or progressive scrub as of the same ecological rank as chief associations like flourishing tracts of woodland.

From the standpoint of succession, the study of the ground vegetation of woodland is a matter of prime importance both to the phytogeographer and to the forester. By such a study, an inkling may be gained of the possible fate of particular tracts of degenerating forest. For example, where the ground vegetation consists of heathy undershrubs, such as bilberry (*Vaccinium Myrtillus*) and heather (*Calluna vulgaris*), and of humus-loving grasses, such as purple moor-grass (*Molinia caerulea*) and silver hair-grass (*Deschampsia flexuosa*), it may often be inferred that the wood is on its way towards heath or moor; whilst where the ground vegetation consists largely of wood soft-grass (*Holcus mollis*), the wood is more likely to be converted ultimately into grassland. On the other hand, a wood whose ground vegetation consists largely of shade-loving species, such as wood anemone (*Anemone nemorosa*) and woodruff (*Asperula odorata*), shows no signs of degeneracy and is probably in a more or less static condition.

Progressive and Retrogressive Scrub

Retrogressive scrub is so-called because it results from the degeneration of chief associations (see p. 21). Progressive scrub, on the other hand, leads to the establishment of chief associations. As has been stated, the latter type of scrub is of uncommon occurrence in this district. However, small examples of progressive scrub may be seen here and there. They are found on fresh soils at the base of cliffs, on projecting ledges of cliffs, on screes, and in deserted quarries. On the sandstones and shales, in fact, they are almost limited to the last kind of locality. Mr Margerison (1907—8) has published an excellent account of the vegetation of sandstone quarries near Bradford, Yorkshire; and his account is of more than local interest. Mr Margerison shows that the plant succession of some of the disused sandstone quarries which he investigated has reached the stage of a birch (chiefly *Betula pubescens*) wood; and it is possible that this stage may in time be replaced by a wood of *Quercus sessiliflora*. On the limestone areas, however, the culminating stage is an ash (*Fraxinus excelsior*) wood. Retrogressive scrub is usually open: progressive scrub is frequently closed, and often forms dense impenetrable tangles of low woody vegetation.

Comparison of the Types of Retrogressive Scrub

The different types of scrub of the district, then, are related to chief associations of woodland. The decay of oak and birch woods results in types of scrub which should be classed in the same plant formation as the oak and birch woods; and the decay of ash woods results in types of scrub which should be placed in the same plant formation as the ash woods. There is, so far as one can judge, little or no essential change in the essential nature of the habitats of the various woods and their related scrub; and intermediate examples are so very numerous and varied that it would seem to be quite impossible to decide on any natural line of demarcation between woodland and related scrub.

In subordinate associations such as these, which are "on the move," *i.e.*, which are kinetic and not static, it is a difficult matter to give really satisfactory lists of plants. If the localities are not very carefully chosen, one finds, in the case of scrub, for example, that one takes a list of species almost characteristic of a wood or a list almost characteristic of grassland. The following lists, however, are taken from typical cases of scrub, though another observer might easily include either more woodland species or more grassland species. As it is, it will be seen there are very few species of the scrub which do not occur either in the woodland or grassland associations; and from this point of view alone, it is not possible to regard the different types of scrub that occur in this district as constituting a natural group of plant communities.

	Scrub on sandstone plateaux and slopes	Scrub on shaly slopes	Scrub on limestone slopes
Trees, shrubs, and undershrubs			
Salix caprea	r	r	r
S. caprea × cinerea	vr	vr	vr
S. cinerea	o	o	o
var. aquatica	r	r	—
var. oleifolia	r	r	r
S. aurita	o	o	—
S. aurita × caprea	r	r	—
S. aurita × cinerea	r	r	—
S. repens	vr	vr	—
Corylus Avellana	r	r	la to s
Betula pubescens	r to a	r to a	—
var. parviflora	r	r	—
forma denudata	o	o	—
Quercus sessiliflora	r to o	r to o	—
Pyrus Malus	r to o	r to o	r to o
P. Aucuparia	o	r	vr

	Scrub on sandstone plateaux and slopes	Scrub on shaly slopes	Scrub on limestone slopes
Trees, shrubs, and undershrubs			
Pyrus Aria	—	—	vr
Crataegus Oxyacantha	la	la	la to s
Rubus fruticosus (agg.)	r to a	r to a	r to a
Rosa canina	r to o	r to o	r to o
"R. mollis"	r	r	r
R. tomentosa	r	r	r
R. spinosissima	—	—	la
Prunus spinosa	r to o	r to o	r to o
Ulex Gallii	r to a	r to o	—
U. europaeus	l	l	—
Empetrum nigrum	r to o	r	—
Ilex Aquifolium	r	r to o	vr
Euonymus europaeus	—	—	r
Rhamnus Frangula	vr	—	—
R. catharticus	—	—	r
Hedera Helix	r	r	o
Cornus sanguinea	—	—	r
Calluna vulgaris	a	r	vr
Erica Tetralix	l	—	—
E. cinerea	lo	r	—
Vaccinium Myrtillus	a	r to o	—
V. Vitis-idaea	la	r	—
Fraxinus excelsior	—	—	r to a
Lonicera Periclymenum	r	r	r
Herbaceous species			
Nephrodium montanum (=N. Oreopteris)	r to o	r to o	—
N. Filix-mas	o	o	o
var. paleaceum	o	o	—
N. aristatum	r to o	r	—
Asplenium spp.	—	—	o
Blechnum spicant	o	r	—
Pteris aquilina	r to a	r to a	r
Rumex Acetosella	r	o	
Lychnis dioica	—	o	o
Stellaria Holostea	—	r to o	r to o
Thalictrum collinum	—	—	l
Anemone nemorosa	—	v	r to o
Ranunculus bulbosus	—	r	o
R. Ficaria	—	r	r to la
Corydalis claviculata	l	l	—
Cardamine impatiens	—	—	r
Sedum acre	—	—	la
Saxifraga hypnoides	—	—	l
S. tridactylites	—	—	la
Rubus saxatilis	—	—	l

	Scrub on sandstone plateaux and slopes	Scrub on shaly slopes	Scrub on limestone slopes
Herbaceous species			
Fragaria vesca	—	r	r to o
Potentilla erecta	o	o	r
Poterium Sanguisorba	—	—	a
Alchemilla filicaulis	—	?	o
Agrimonia Eupatoria	—	—	o
Lathyrus montanus	r	r to o	r
Geranium sanguineum	—	—	l
Oxalis Acetosella	r	r	r
Polygala vulgaris	—	r	r to a
P. serpyllacea	r	r	
Mercurialis perennis	—	r	r to a
Hypericum pulchrum	r	r	r
H. perforatum	—	r	r to o
H. hirsutum	—	—	r to la
H. montanum	—	—	l
Helianthemum Chamaecistus	—	—	a
Viola hirta	—	—	o
V. Riviniana	r	r to o	r to o
Epilobium montanum	—	r	r
E. angustifolium	r	r	r
Conopodium majus	—	r to o	r to o
Heracleum Sphondylium	—	r	o
Daucus Carota	—	—	r
Pyrola media	vr	—	—
P. minor	vr	—	—
Primula veris	—	—	o
P. vulgaris	vr	vr	r to o
Gentiana Amarella	—	l	o
Teucrium Scorodonia	o	r to o	r
Scutellaria minor	l	—	—
Nepeta hederacea	—	l	la
Origanum vulgare	—	—	o to a
Thymus Serpyllum (agg.)	—	—	o to a
Veronica officinalis	—	r to o	o
Digitalis purpurea	o	o	—
Melampyrum pratense (agg.)	r	r	—
Plantago media	—	—	o
P. lanceolata	—	r to o	o
Galium verum	—	—	a
G. Cruciata	—	l	r to o
G. saxatile	o to a	a	—
G. sylvestre	—	—	la
Scabiosa Succisa	r	r	r
S. Columbaria	—	—	r
Campanula Trachelium	—	—	r
C. rotundifolia	r	r to o	r
Jasione montana	r	r	—

	Scrub on sandstone plateaux and slopes	Scrub on shaly slopes	Scrub on limestone slopes
Herbaceous species			
Solidago Virgaurea	o	—	r to o
Antennaria dioica	r	—	—
Gnaphalium sylvaticum	vr	vr	—
Inula squarrosa	—	—	r
Chrysanthemum Leucanthemum	—	r	r to a
Senecio sylvaticus	r	—	—
Carlina vulgaris	vr	vr	r to o
Carduus nutans	—	—	r to o
Cnicus eriophorus	—	—	vr
C. palustris	o	o	o
Centaurea Scabiosa	—	—	r to o
Picris hieracioides	—	—	r
Leontodon hispidus	—	r	o
Hieracium spp.	r	r	r to la
Taraxacum officinale (agg.)	—	r	—
Anthoxanthum odoratum	—	o	o
Agrostis tenuis (=A. vulgaris)	r	la	la
Holcus mollis	l	r to o	—
Deschampsia flexuosa	la	la	—
Avena spp.	—	—	r to o
Arrhenatherum elatius	—	vr, l	a
Triodea decumbens	r	r to o	r
Molinia caerulea	la	r	—
Koeleria cristata (agg.)	—	—	o
Melica nutans	—	—	vr
Briza media	—	r	o
Festuca ovina	r to o	la	la
Brachypodium sylvaticum	—	r	r to a
Carex binervis	r to o	o	—
C. pilulifera	r	o	r
C. ornithopoda	—	—	l
Juncus spp.	la	la	r
Luzula multiflora	o to a	o to a	—
forma congesta	o	o	—
L. campestris	—	la	a
L. pilosa	r	r to o	r to o
Convallaria majalis	—	—	l
Scilla non-scripta	—	r	vr
Listera ovata	—	vr	r to o
Helleborine atrorubens	—	—	l
Habenaria viridis	—	—	r
"Ophyrs apifera"	—	—	vr
Orchis mascula	—	r	r to la
"O. pyramidalis"	—	—	vr

CHAPTER IV

GRASSLAND ASSOCIATIONS

Distribution of the grassland. Types of grassland. I. Grassland of the sandstones and shales: siliceous grassland; (1) Nardus grassland; mixed siliceous grassland; (2) Molinia grassland. Relationships of the plant associations of the siliceous soils. II. Grassland of the Limestone: calcareous grassland: mixed calcareous grassland; transitional calcareous grassland. Calcareous heath. Pseudo-calcareous heaths. Species of the calcareous grassland and the siliceous grassland. Relationships of the plant associations of the siliceous and the calcareous soils.

Distribution of the Grassland

As is the case with woodland and scrub, grassland occurs, in general, on the slopes of the hills. Where the hill-slopes below about 1500 feet (457 m.) are not cultivated and not occupied with woodland or scrub, there natural or uncultivated grassland prevails. The cultivated grassland or permanent pasture is dealt with in Chapter VIII. On the whole, natural grassland is more extensive on the limestones than on the sandstones and shales; and, with regard to the non-calcareous soils, it is, in proportion to their extent, much more extensive on the shales than on the sandstones.

At the present time, although grassland ascends to higher altitudes than the woodland, it is rather rare at altitudes above the present limit of scrub. It is highly probable that almost all the present grassland—both natural and cultivated—was once wooded, and that even now it is almost all capable of being successfully reafforested (see Chapter VIII).

In a few places, however, as on the elevated summit of Bleaklow Hill, at a height of about 2000 feet (610 m.), sub-Alpine grassland occurs on ground which has probably never been tree-clad—at least, not in post-Tertiary times. It will be shown later on that such sub-Alpine grassland occurs, so far as this district is concerned, on sites which were comparatively recently covered with peat; and the peat having suffered denudation, plants of the siliceous pasture have successfully invaded areas which were once peat-clad.

Natural grassland is rather uncommon on the less elevated plateaux, for these are usually either occupied by moorland associations or they are under cultivation.

TYPES OF GRASSLAND

Two main types of grassland occur in the district. One is developed on the siliceous soils, the other on the calcareous soils. The former type of grassland is characterized by the great abundance of heath-loving or humus-loving species, and is termed *siliceous grassland*. The non-calcareous or siliceous soils allow of the formation and accumulation of acidic humus in the soil; but any great excess of this is, on steep slopes, prevented by the denuding action of rain and melting snow. Instead of the accumulation of peat, we get, on steep slopes, a slow but continuous exposing of new soils. Such conditions favour the growth of sward-forming grasses rather than of heathy under-shrubs, for although newly exposed siliceous soils are much poorer in soluble minerals than calcareous soils, they are richer than sour peaty soils. On the plateaux, however, the acidic humus or peat may accumulate; and the ground is then invaded by heather (*Calluna vulgaris*) and ecologically allied species. It seems certain that the steep shaly slopes will never become peat clad, whilst the grassland of the non-calcareous plateaux will probably be ultimately converted into moorland.

The sub-Alpine pasture above mentioned is essentially identical in its ecological and floristic characteristics with siliceous pasture; but, as it occurs at higher altitudes, it has fewer associated species. All the species, however, which actually occur on the sub-Alpine pasture, occur on heath pasture also; and the two associations therefore are placed

 W. B. Crump

Figure 14.

Siliceous Grassland.

Blue moor-grass (*Molinia caerulea*) in the foreground. Mat-grass (*Nardus stricta*) and silver Hair-grass (*Deschampsia flexuosa*) covering the whole of the slopes and summit of the hill.

in the same plant formation. On the Pennines further north (see Smith and Rankin, 1903: 154), similar sub-Alpine pasture occurs; but there one species, *Poa alpina*, occurs which has not been found in the Peak District. Sub-Alpine pasture, characteristic of the Scottish mountains, has been described by R. Smith (1900 *b*: 454).

On the calcareous soils, the grassland is poor in heath-loving or humus-loving species but rich in lime-loving species, and this association is termed *calcareous grassland.* A certain number of species (see the lists of plants given later on in the chapter) are common to siliceous grassland and calcareous grassland. The two types of grassland are related, directly or indirectly, to the presence or absence of calcium carbonate in the soil. On the limestones, it is only at the higher altitudes, where the soils are leached by rain and therefore contain much less lime, that calcareous grassland approaches siliceous grassland in its ecological and floristic characteristics. On the lower slopes of the calcareous hills where the soil is rich in lime, the acidic humus which favours the growth of the plants of the siliceous grassland does not appear to be formed; and it certainly does not accumulate.

A certain amount of grazing of sheep and cattle takes place on many parts of the grassland; but the amount is, on the whole, rather small. The land is not artificially manured or drained. On the sub-Alpine grassland, no grazing or manuring takes place at all.

Other types of grassland occur in other parts of the country, more especially in central and southern England. Clayey and fresh loamy soils, for example, are characterized by the absence of both humus-loving and of lime-loving species; and the grassland of such soils may therefore be termed *neutral grassland.* A fourth type occurs on the flat lands which occur near rivers and which are liable to periodical inundations: this may be termed *alluvial or fen grassland.* An analysis of the grasslands of Orkney has recently been published (Scarth, 1911).

"Permanent pasture" is an agricultural term in use in this country to denote grazing land which has, in general, been ploughed up at least once, and which is artificially manured (see Chapter VIII).

I. GRASSLAND OF THE SANDSTONES AND SHALES: SILICEOUS GRASSLAND

Two types of siliceous grassland have been described in previous botanical surveys of the Pennines (Smith and Moss, 1903: 384; Smith and Rankin, 1903: 158; Lewis 1904 *a*: 323, 1904 *b*: 275), and have been distinguished as wet and dry. The most abundant and characteristic grass of the drier type of siliceous grassland is the mat-grass (*Nardus stricta*) and that of the wetter type is the purple moor-grass (*Molinia caerulea*) (cf. figure 14). The two species are respectively dominant in the two associations since they nearly monopolize the ground and form the great bulk of the turf, the associated species being therefore more or less controlled by them. The former association may therefore be termed Nardus grassland (*Nardetum strictae*) and the latter Molinia grassland (*Moliniëtum caeruleae*). To some extent, the associations are layered plant communities; and the smaller plants receive a certain amount of shade and shelter from the dominant ones. As is usual in plant associations, one or another of the dependent species occasionally becomes more or less social; and thus plant societies and facies arise.

(1) Nardus Grassland

Typical Nardus grassland (see figure 13) occurs on steep shaly slopes of the non-calcareous hills. In summer, this association is characterized by a grassy turf, grey-green in colour, dry and slippery. In late autumn, winter, and early spring, the ground is damp and sodden; and the bleached haulms of the mat-grass (*Nardus stricta*) give tone to the landscape, and may be recognised at a considerable distance. The silver hair-grass (*Deschampsia flexuosa*) is, in this district, a constant associate. In winter, the mat-grass is very much more conspicuous than the hair-grass, as, during this season, the short leaves of the latter are usually more or less hidden beneath the long, white sprays of the dead shoots of the former. Under such conditions, the hair-grass, even though very abundant, is apt to be overlooked. It is only in early summer, when the tall, purple scapes of the hair-grass are in bud, flower, or fruit that this species becomes obtrusively conspicuous; and, at such times, it gives the tone and colour to the whole association. There seems little doubt, however, that the Nardus association of the Peak District is ecologically identical with that of the Wicklow Hills (Pethybridge and Praeger, 1905: 157) and that of the northern Pennines (Lewis, 1904 *a*: 324; 1904 *b*: 275), even though the silver hair-grass is not included in the lists of these districts.

The silver hair-grass of the hills of the Peak District belongs to the form with short, wiry, and sub-squarrose leaves (*Deschampsia flexuosa*, ? var. *montana*): the form in the oak and birch woods has much longer, more limp, and more slender leaves. Woodhead (1906: 383) has described and figured the structural differences of some of the forms of this plant.

The two grasses (*Nardus stricta* and *Deschampsia flexuosa*) remain co-dominant up to the edge of the moorland plateau, which frequently occurs at about 1500 feet (457 m.). Below about 1250 feet (381 m.), the common bent-grass (*Agrostis vulgaris*) is often an abundant associate, giving rise to a distinct facies. In the late summer months, its delicate and purple panicles colour the hill sides. As lower altitudes are approached, this species becomes increasingly abundant at the expense of the mat-grass (cf. page 112). The sheep's fescue-grass (*Festuca ovina*) is also often associated; and this species sometimes forms plant societies and facies.

The shaly hill-slopes of the Pendleside (or Yoredale) series which encircle the upper Edale valley afford an extensive and continuous expanse of Nardus pasture. On the north of this upland valley are the slopes of the Peak, on the south the slopes of the Mam Tor range, and on the east the slopes of the Colborne moors. Such a great expanse of Nardus grassland is not seen elsewhere in the district. In the sheltered Grindsbrook clough, the bracken (*Pteris aquilina*) asserts itself very strongly: the dwarf furze (*Ulex Gallii*) occurs in small patches here and there; and the springs of water on the hill sides are marked by clumps of the common rush (*Juncus effusus*).

The last three species give to the association very different aspects or facies. The bracken, where the soil is dry and the locality sheltered, sometimes occurs in extensive sheets (see figure 15). The gorse (*Ulex Gallii*) is never very prominent in this district, like it is, for example, on the Malvern Hills or on the Wicklow Hills (cf. Pethybridge and Praeger, 1905: 153, plates 7 and 8); but it occurs in patches in dry and fairly exposed localities. The rush (*Juncus effusus* and *J. effusus* forma *compactus*), in damp places, and independently of conditions of shelter or exposure, is an abundant and characteristic associate. The bracken and the rush, in fact, are, in many places harvested by the upland farmers (see figures 15 and 16).

The relationships of the various facies and aspects of the Nardus association may be indicated diagrammatically as follows:—

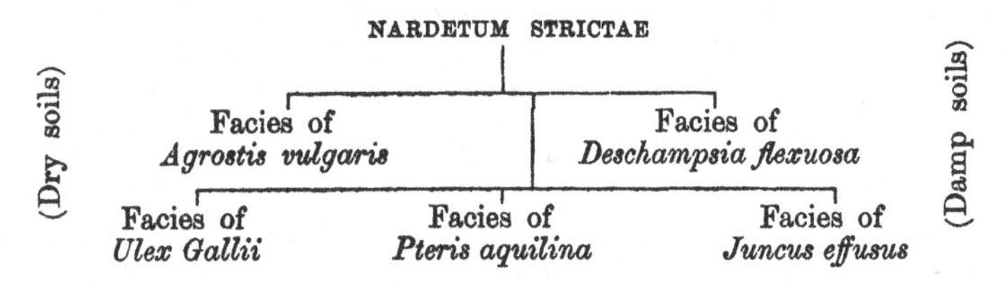

The bracken facies (see figure 15) and the rush facies (see figure 16) are very well developed in this district. For example, in ascending Longdendale (above Glossop), one meets with Nardus grassland on the hill-slopes. The bracken here varies from being a rare member of the association to a subdominant member. However, the general habitat conditions seem so very constant, and the bracken itself so very variable in amount in different parts of the association (even within quite small areas), that it seems impossible to regard the bracken as modifying the association any more than producing a facies. Towards the head of Longdendale, between Woodhead and Dunford Bridge, the bracken becomes less general; whilst the rush becomes a normal and an abundant member of the Nardus grassland. Such places are always ill-drained; and from them, the gorse and the bracken are invariably absent. The cotton-grass (*Eriophorum angustifolium*) sometimes occurs. Such a Juncus facies of siliceous grassland must be distinguished from the Juncus swamp (Juncetum effusi) described in Chapter VI.

The bracken is one of the most accommodating of plants as regards its requirements. It is present in nearly all woods, except in the shadiest, dampest, and most calcareous places;

Copyright *W. B. Crump*

Figure 15.

Siliceous grassland.

Facies of Bracken (*Pteris aquilina*). Stack of bracken litter in the foreground. The trees are Hawthorns (*Crataegus Oxyacantha* = *C. monogyna*).

and it is probable that the present upper altitudinal limits of the bracken approximate very closely with the upper altitudinal limits of the ancient and pre-historical forests. Sometimes the bracken spreads from the Nardus grassland into the adjoining moorland associations, not uncommonly into the heather (*Calluna vulgaris*) moor when this occurs in rather sheltered valleys and depressions, and rarely into the cotton-grass (*Eriophorum vaginatum*) moor when the peat of this is becoming drier. In the Peak District, the bracken commonly ascends to 1500 feet (457 m.) or 1550 feet (472 m.), above which altitudes it becomes local and rare. Woodhead (1906: 360) states that it ascends to 1700 feet (518 m.); but in England it is quite exceptional to meet with the plant at such an altitude.

Pethybridge and Praeger (1905: 155) describe areas of bracken in the district south of Dublin. The list of species which these authors give indicates that species characteristic of Nardus grassland are numerous in such areas. The authors state that in the month of May, the bracken areas often exhibit glorious sheets of blue, white, and yellow due to the abundance of the blue-bell (*Scilla non-scripta*), violet (*Viola Riviniana*), speedwell (*Veronica Chamaedrys*), earth-nut (*Conopodium majus*), lesser celandine (*Ranunculus Ficaria*), and primrose (*Primula vulgaris*). Such a display of flowers is unknown in connection with the bracken areas of the upland slopes of the Pennines, but may occasionally be seen as the lowlands are reached.

Pethybridge and Praeger (1905: 150 and 153) map two plant communities of gorse. In one, at lower altitudes, *Ulex europaeus* is conspicuous; and in the other, at higher elevations, *Ulex Gallii* is exceedingly abundant. In this district, the two species have similar altitudinal relationships; but they are not specially abundant, especially *U. europaeus*. In Somerset, this altitudinal relationship perhaps does not hold good, for the two species frequently occur side by side. In certain localities in the south of England, a third species of gorse (*Ulex minor* = *U. nanus*) becomes locally a very important element on heaths and heathy grasslands.

Ostenfeld (1908: 966) has described a grass-slope "formation" of the Faeröes; and this would appear to be very closely related to the siliceous grassland of Britain, as about eighty per cent. of the species are common to the two types of vegetation.

Brockmann-Jerosch (1907: 248, etc.) has described an association (*Bestandetypus*) of Nardus stricta in the Puschlav region of the Swiss Alps. This Alpine Nardetum is found at much higher altitudes than occur anywhere in England, and contains many species which are not indigenous in this country: still, about thirty per cent. of the species given by Brockmann (1907: 305—8) are found in the Nardetum of the Peak District of Derbyshire; and probably the two associations should be placed in the same plant federation (see Moss, 1911), but in different geographical plant formations. The elements common to the two associations are the following:—

Botrychium Lunaria	Achillaea Millefolium
Rumex Acetosella	Hieracium Pilosella
Ranunculus acris	Anthoxanthum odoratum
Trifolium repens	Deschampsia flexuosa
T. pratense	Sieglingia decumbens
Vaccinium Vitis-idaea	Nardus stricta
Calluna vulgaris	Carex caryophyllea
Antennaria dioica	Luzula campestris

It has already been emphasised that the woods of the district gradually pass into scrub; and similarly the scrub imperceptibly passes into grassland. Just as there are many localities which are difficult to determine either as woodland or scrub, so there are many other places which are difficult to determine either as scrub or grassland. Again, whilst much of the scrub represents degenerate woodland, much of the grassland represents a still later stage of degeneracy than the scrub. There can be little doubt that the hill-slopes now characterized by grassland were, generally speaking, once wooded; and it is highly probable that most of these slopes are capable of being successfully re-afforested. In the Nardus grassland on the hill-slopes of this district, the following woodland species, among others, still linger here and there:—

Pteris aquilina	Betula pubescens
Nephrodium montanum	Lychnis dioica
N. aristatum	Corydalis claviculata
N. Filix-mas	Oxalis Acetosella
Athyrium Filix-foemina	Geranium Robertianum
Quercus sessiliflora	Ilex Aquifolium

Copyright *W. B. Crump*

Figure 16.

Siliceous grassland.

Facies of the common Rush (*Juncus effusus*).

Harvesting the rushes.

Pyrus Aucuparia	Ajuga reptans
Crataegus Oxyacantha	Digitalis purpurea
Lathyrus montanus	Melampyrum pratense (agg.)
Vicia sepium	Holcus mollis
Viola Riviniana	Luzula pilosa
Conopodium denudatum	Orchis mascula
Heracleum Sphondylium	Scilla non-scripta

Whilst opinions may differ as to whether or not the grassland just described is wholly or only in part due to man's interference, it appears to be generally accepted that such tracts were formerly clothed with forest; and Warming (1909: 326) even goes so far as to say that "were the human race to die out," the grasslands of the lowlands of north Europe "would once more be seized by forest, just as their soil was originally stolen from forest." As regards the Nardus grassland of the hill-slopes of this district, it seems incontestable that it is an association which has, on the whole, resulted from the degeneration of oak and birch woods. The fundamental conditions of the habitat have been but slightly altered in the process; and, therefore, the oak and birch woods, the Nardus grassland, and the various transitional stages of scrub are placed in one and the same plant formation.

The following is a list of the more typical and abundant plants which occur in the Nardus grassland: full lists of grassland species are given at the end of the chapter:—

Dominant species

Nardus stricta

Sub-dominant species

Deschampsia flexuosa

Locally sub-dominant species

Pteris aquilina

Locally abundant species

Ulex Gallii	Festuca ovina
Agrostis vulgaris	Juncus effusus

Less abundant and rarer species

Botrychium Lunaria (local)	Potentilla erecta
Blechnum spicant	Lotus corniculatus
Rumex Acetosella	Cytisus scoparius

Lathyrus montana
Polygala serpyllacea
Hypericum pulchrum
Viola lutea (local)
Veronica officinalis
Euphrasia officinalis (agg.)
Plantago lanceolata
Galium saxatile
Campanula rotundifolia
Crepis virens
Hieracium Pilosella
Anthoxanthum odoratum
Briza media
Carex flacca
C. binervis
C. pilulifera
Juncus squarrosus
Luzula erecta

Mixed Siliceous Grassland

Bordering on the upper limits of the cultivated land on rough escarpments of the cloughs, and on the outskirts of woods, a type of siliceous pasture occurs which is much richer in associated species than the Nardetum just described. The type of siliceous pasture here referred to is found usually at altitudes below 1000 feet (305 m.), is frequently fenced, and, to some extent, is grazed over by cattle; but it is probable that it has never been ploughed or drained. Many of the species are of local occurrence, and probably represent the remains of a primitive flora which flourished in the open spaces of the aboriginal woods at comparatively low altitudes. The flora of this association has been given by Crump (1904: xli) and by Smith and Moss (1903: 385), so far as the Leeds and Halifax district is concerned.

The following list illustrates the wealth in species of this type of siliceous grassland:—

Botrychium Lunaria	lo	Quercus sessiliflora (dwarfed)	l
Ophioglossum vulgatum	la	Rumex Acetosa	o
Pteris aquilina	r to la	R. Acetosella	la
Blechnum spicant	o	Ranunculus acris	o
Nephrodium montanum	lo	R. bulbosus	r to o
Salix caprea	r to o	Cerastium vulgatum	l
S. caprea × cinerea	r	Stellaria graminea	o
S. aurita	lo to la	Prunus spinosa	la
S. aurita × caprea	r	Rubus fruticosus (agg.)	la
S. aurita × cinerea	r to lo	Potentilla erecta	o to a
S. cinerea	la	P. procumbens	r
S. repens	r	P. erecta × procumbens	r
Betula pubescens (dwarfed)	l	Alchemilla vulgaris (agg.)	r to o
forma denudata (dwarfed)	l	Rosa canina	o
Corylus Avellana	l	R. tomentosa	r

Pyrus Malus	r to o	Bellis perennis	o
P. Aucuparia (dwarfed)	r	Gnaphalium sylvaticum	vr
Crataegus Oxyacantha	r to a	Achillaea Millefolium	o
Genista anglica	r	Chrysanthemum Leucanthemum	o
G. tinctoria	r, la	Senecio Jacobaea	la
Ulex Gallii	la	Carlina vulgaris	vr
U. europaeus	r, la	Centaurea nigra	a
Ononis repens	r	Crepis virens	o to a
Trifolium medium	r to o	Hypochaeris radicata	o
T. pratense	o	Leontodon hispidus	o
T. repens	o	L. autumnalis	a
Lotus corniculatus	a	Hieracium Pilosella	o to a
Vicia angustifolia	r	H. vulgatum	r
Lathyrus pratensis	o	H. boreale	r
L. montanus	o	Taraxacum officinale	o
Linum catharticum	a	*var.* maculiferum	r
Polygala serpyllacea	o	Anthoxanthum odoratum	o
Hypericum humifusum	lr	Agrostis tenuis	a to s
H. pulchrum	o	Deschampsia caespitosa	r to o
Viola lutea	r, la	D. flexuosa	o to a
V. Riviniana	r, la	Holcus mollis	l
Empetrum nigrum	r	H. lanatus	o
Ilex Aquifolium (dwarfed)	o	Arrhenatherum avenaceum	l
Pimpinella Saxifraga	o	Sieglingia decumbens	o
Conopodium majus	o	Cynosurus cristatus	l
Vaccinium Myrtillus	r to a	Molinia caerulea	l
V. Vitis-idaea	r	Briza media	o
Calluna vulgaris	r to a	Festuca ovina	o to a
Erica cinerea	r to a	F. duriuscula	r to a
Primula veris	r	Brachypodium sylvaticum	l
Centaureum umbellatum	r	P. vulgaris	r
Gentiana Amarella	la	Carex ovalis	l
G. baltica	vr	C. flacca	r to o
Thymus Serpyllum (agg.)	o to r	C. pilulifera	r to o
Stachys officinalis	o	C. caryophyllea	l
Teucrium Scorodonia	r to la	C. hirta	l
Digitalis purpurea	o	C. pallescens	r
Veronica officinalis	o	C. binervis	l
V. Chamaedrys	la	Luzula pilosa	lo
Prunella vulgaris	o	L. campestris	la
Euphrasia officinalis (agg.)	a	L. erecta	lo
Rhinanthus Crista-galli	r, la	Listera ovata	r
Plantago lanceolata	a	Orchis maculata	?
Galium saxatile	a	O. ericetorum	lo
Scabiosa Succisa	o	Habenaria conopsea	vr
S. arvensis	r	H. viridis	r
Campanula rotundifolium	o to a	H. bifolia	vr
Jasione montana	r to o	H. chloroleuca	vr
Solidago Virgaurea	o		

(2) Molinia Grassland

The wetter type of siliceous grassland (cf. p. 106), dominated by the purple moor-grass (*Molinia caerulea*), is of far less extent in this district than the Nardus grassland, and much more local in its occurrence.

The Molinia grassland occurs, as a rule, on flatter ground than the Nardus grassland (cf. figure 14). In a very general way, the Molinia grassland affects the ground overlying the sandstone rocks and the Nardus grassland the steep slopes of the shales. Occasionally, the Molinia grassland occurs on shales where the drainage is obstructed by boulders which have fallen from an escarpment of sandstone. Invariably, the soil of the Molinia grassland is wet, often very wet, and more or less peaty. Such soil is, in this district, always acidic; but Molinia is by no means always confined to acidic soils. In East Anglia, for example, Molinia occurs on alkaline peaty soils; and here the associated species are different from those of acidic soils. On acidic soils, the Molinia grassland is transitional between grassland and moorland, as was recognised by R. Smith (1900 *b*: 454); and examples occur which might quite fairly be placed among the moorland associations. The moorland character is seen in its acidic peaty soil, often supersaturated with moisture, and in the abundance of associated species which characterize certain parts of the moorland. Some of the Molinia associations of this district are almost demonstrably derived from oak or birch woods with a heathy ground flora; and there can be little doubt that it sometimes develops into moorland. Transitions of this nature occur on the south side of Longdendale, near Crowden railway station. In a few cases, perhaps, Molinia invades the Nardus grassland; and Molinia is frequently an associate in the heather moor.

The plant (*Molinia caerulea*) has a wide range of form and habitat. The variety of form known as *Molinia caerulea* var. *depauperata*, with one-flowered spikelets and shorter leaves and shoots, occurs in very wet places. Usually, the plant (*Molinia caerulea*) is about half a metre high; but here and there a variety (*M. caerulea* var. *major*) with wide-spreading branches of the inflorescence occurs. The plant is deserving of a careful study on account of its wide range of forms and the different nature of its habitats.

A list of the members of the Molinia grassland is appended, and the number of associated species characteristic of the moorland formation is obvious:—

Polytrichum commune	la	Nardus stricta	la
Sphagnum spp.	la	Scirpus caespitosus	r to o
"Lycopodium inundatum"	vr	Eriophorum vaginatum	r to o
Ranunculus Flammula	la	E. angustifolium	la
forma radicans	r to a	Carex curta	r to o
Drosera rotundifolia	r	C. echinata	la
Viola palustris	r to o	C. Goodenowii	o to a
Empetrum nigrum	la	*var.* juncella	r
Hydrocotyle vulgaris	la	C. glauca	o
Andromeda Polifolia	r	C. panicea	o to a
Erica Tetralix	la	Deschampsia flexuosa	r to a
Calluna vulgaris	la	Carex flava	la
Vaccinium Myrtillus	la	*var.* minor	la
V. Oxycoccus	r to a	Juncus effusus	la
Pinguicula vulgaris	r	J. squarrosus	la
Taraxacum palustre	r	J. acutiflorus	la
Agrostis canina	l	Narthecium ossifragum	r to la
Molinia caerulea	a to d	Orchis maculata	?
var. depauperata	r to d	O. ericetorum	r to o
var. major	l		

Relationships of the Plant Associations of the Siliceous Soils

The relationships of the plant associations of the sandstones and shales of the southern Pennines are summarized in the following table:—

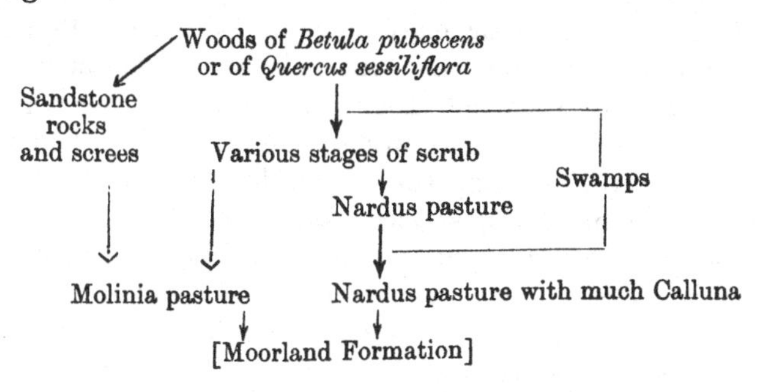

II. Grassland of the Limestone: Calcareous Grassland

In previous works on British plant geography, this group of associations has been variously termed "natural pasture," "limestone hill pastures," and "Permian common" (cf. Smith and Moss, 1903; Smith and Rankin, 1903; Lewis, 1904 *a* and *b*; and Moss, 1907 *a*); but the name calcareous grassland is now becoming general.

Typical calcareous grassland consists of short, grassy turf, largely composed of the sub-aërial parts of the sheep's fescue-grass (*Festuca ovina*). In this district, calcareous grassland is found abundantly on the uncultivated, steep slopes of the limestone dales (see figures 12 and 16). On the limestone plateaux, the soil is frequently leached, and then certain plants of the siliceous grassland enter the association.

It will be seen that the boundaries of the associations of the soils of the sandstones and shales and of the soils of the limestones do not follow any boundaries shown on the geological maps. The latter indicate the boundaries and extent of the subjacent rocks, but do not attempt to deal with the surface soils which alone are related directly to the flora and vegetation. This distinction between the subjacent rocks and the surface soils is adequately emphasized in the treatment of plant formations of the British Isles by Tansley (1911, *passim*).

With slight differences in floristic composition as are indicated in the preceding paragraph, calcareous grassland occurs on all the uncultivated tracts of the various calcareous soils of the country, such as on the chalky boulder clay, the chalk rock and marl, the Jurassic marls and limestones, the Permian or Magnesium Limestone, and the Palaeozoic limestones.

Calcareous grassland is a plant association, or perhaps a group of closely allied associations, characterized by the presence of numerous lime-loving species and by the absence of heath-loving or humus-loving species. Lime-loving species have been variously termed xerophiles, calciphiles, and calcicoles: lime-shunning species have been termed hygrophiles, calciphobes, and silicicoles. It seems highly desirable to subdivide the lime-avoiding species into three classes:—(1) plants of acidic peat, (2) plants of siliceous soils, and (3) plants of sandy soils.

The soil of the calcareous grassland of the limestone slopes is shallow, sometimes not even an inch (about 2·5 cms.) deep. In colour, it varies from a whitish grey when the lime-content is very high, to brownish or even reddish-brown when the lime-content is lower and the iron-content higher. At the foot of a cliff, the soil may be a metre or more in depth: commonly it is about a sixth to a third of a metre deep. The soil of limestones is usually described as being very dry and porous; but it is only the newer and whiter soil of which this may correctly be stated. The older and darker soil is of a marly nature, and is neither specially dry nor specially porous. Similarly, the newer and yellower soil of the sandstones may be dry and porous, whilst the older and blacker soil of the sandstones is retentive of water. Analyses prove that the range of variation of water-content of the soils derived from the limestone rocks is roughly paralleled by that of the soils of the sandstones and shales. Just as the water-content of the non-calcareous soils varies directly as the humus-content, so the water-content of the limestone soils varies inversely as the lime-content. The former result is doubtless due to the water-absorbing properties of humus: the latter seems to be due to the fact that calcium carbonate is dissolved by water containing carbon dioxide; and thus as limestone soils become older they lose more and more lime and acidic humus then tends to accumulate. Marshy places occur on the limestone just as they do on the sandstones and shales. However, it may be said that, in general, such localities are least frequent on the limestones, more frequent on the sandstones, and very numerous on the shales. The marshy places on the limestones bear a very different flora from those of the sandstones and shales, just as the dry limestone soils possess a very different flora from such soils on the sandstones; and it thus appears to be quite impossible to explain the distribution of the humus-loving and the lime-loving species respectively by any relations of the water-content.

The abundance of the bracken (*Pteris aquilina*), the gorse (*Ulex Gallii*), and the rush (*Juncus effusus*), which is so very noticeable a feature of the various types of siliceous grassland, is not seen on the calcareous grassland of this district. In fact, on the limestone slopes below 1000 feet (305 m.), these plants are absent or rare; and even on the more or less leached soils above that altitude and on the limestone plateaux, all three species are only of local occurrence. Thus the calcareous grassland presents a different physiognomy from much of the siliceous grassland, the former having a cleaner aspect and a greener and more regular turf. In addition to the above gregarious plants, many other humus-loving species are absent or almost absent from the calcareous grassland of the slopes of the limestone dales; and the following is a list of such species, omitting the plants of marshy places:—

Pteris aquilina
Nephrodium montanum
(=N. Oreopteris)
Athyrium Filix-foemina
Salix repens
S. aurita
Cytisus scoparius
Genista anglica
G. tinctoria
Ononis repens
Lathyrus montanus
Polygala serpyllacea
Empetrum nigrum
Hypericum humifusum
Spergularia rubra
Potentilla erecta
P. procumbens
Ulex Gallii
U. europaeus
Calluna vulgaris
Erica cinerea
E. Tetralix
Vaccinium Myrtillus
V. Vitis-idaea
Scutellaria minor
Digitalis purpurea
Melampyrum pratense
Galium saxatile
Scabiosa Succisa
Jasione montana
Gnaphalium sylvaticum
Holcus mollis
Aira praecox
Deschampsia flexuosa
Molinia caerulea
Nardus stricta
Carex binervis
C. Goodenowii
var. juncella
Juncus squarrosus
Luzula multiflora
forma congesta
Orchis ericetorum

On the other hand, the following plants, excluding those of the marshy places (see Chapter VI), are found in some parts of the calcareous grassland but are absent or quite rare in those of siliceous grassland:—

1. In grassy places:—

Sedum acre
Spiraea Filipendula
"Potentilla verna"
Agrimonia Eupatoria
Poterium Sanguisorba
Anthyllus Vulneraria
"Trifolium striatum"
T. filiforme
Hippocrepis comosa
Hypericum hirsutum
Viola hirta (agg.)
Daucus Carota
Satureia Clinopodium
Origanum vulgare

Verbascum Thapsus
Plantago media
Asperula cynanchica
Galium sylvestre
Scabiosa Columbaria
Campanula glomerata
Pulicaria dysenterica
Inula squarrosa
Senecio erucifolius
Picris hieracioides
Leontodon hirtus
Hieracium spp.
Avena pratensis
A. pubescens
Koeleria cristata (agg.)
"Bromus erectus"
Brachypodium pinnatum
Carex ornithopoda
Ophrys apifera
"O. muscifera"
Orchis morio
O. ustulata
O. pyramidalis

2. In rocky places, particularly where sheltered:—

Asplenium viride
A. Trichomones
A. Adiantum-nigrum
A. Ruta-muraria
Cystopteris fragilis
Thalictrum minus
Draba muralis
Sedum Telephium
"Saxifraga sphonhemica"
S. hypnoides
Rosa spinosissima
Geranium lucidum
G. sanguineum
Hypericum montanum
Pimpinella magna
Satureia Acinos
Galium asperum
Valerianella carinata
V. olitoria
Centaurea Scabiosa
Allium vineale
A. oleraceum

3. In places, where the soil is loose, all semi-ruderal plants, occurring, as a rule, most abundantly on the refuse-heaps ("rakes") of old lead-mines or modern gravel-workings:—

Arenaria verna
A. serpyllifolia
Cardamine hirsuta
Cochlearia alpina
Hutchinsea petraea
Arabis hirsuta
Sisymbrium Thalianum
Erophila verna
var. virescens
E. praecox
"E. inflata"
Thlaspi virens
"T. sylvestre"
Saxifraga tridactylites
Alchemilla arvensis
Viola lutea
var. amoena
V. calaminaria
Myosotis collina
Sherardia arvensis
Carduus nutans
Cnicus eriophorus

From the above lists, it will be seen that calcareous grassland differs greatly from siliceous grassland not only in the soil conditions but also in the floristic composition. In my judgment, the edaphic characteristics of the habitats are so essentially different that the two types of grassland, siliceous grassland and calcareous grassland, should not be placed in the same plant formation; and the same conclusion is indicated by a study of their floristic composition and the related plant associations.

Gradmann (1909: 94) has maintained that a plant formation can be defined floristically; and if this conclusion be accepted, it would seem to be indicated that the siliceous grassland and the calcareous pasture, although often possessing the same physiognomy and the same plant form, must be assigned to different plant formations.

On bushy banks, where there is some shelter from the wind and where the soil is comparatively damp, many shrubs of the ash woods and of calcareous scrub occur; and these shrubs, in their turn, shelter several herbaceous species of the ash woods and scrub. Many of such communities, in fact, appear to be progressive associations which will finally become ash woods; and it is impossible to draw any hard-and-fast boundary line between woods, scrub, and grassland either of the siliceous or the calcareous soils of this district. The transitional associations of these hill-slopes are strictly analogous with the transitional associations occurring on the wet, acidic, peaty soils of the "Hochmoors so abundantly scattered in the foothills on the Jura ridges, the Black Forest, and the Vosges. In contrast to the Hochmoors of the north German plain, there occurs here, as is well known, *Pinus montana* in great communities, but by no means everywhere: even on the moors where it flourishes, wide stretches are often quite free from it. If one now starts with the ordinary physiognomical division [of forests, scrub, grassland, etc.], it becomes necessary to split the natural and sharply defined plant community of the Hochmoor into at least three if not into four or five formations; and these must further be assigned to the most varied positions in the system. According as *Pinus montana* forms well-developed trees, or is the dwarfed form, or is absent altogether, the bit of the Hochmoor in question belongs to the forest formations, or to the scrub formations, or to the moss formation: where Ericaceae occur socially the same Hochmoor becomes a dwarf-shrub formation: where a turf of Carices, Eriophorum, or Scheuchzeria predominate, we have a 'meadow.' And yet the floristic composition is almost exactly the same: the local conditions,

Copyright *W. B. Crump*

Figure 17.

Calcareous Scrub and Grassland.
Rocky hill slope of Carboniferous Limestone.

the ecological relations show scarcely any perceptible alteration; and the *soi-disant* 'formations' everywhere pass imperceptibly one into another" (Gradmann, 1909: 92). This criticism of a method of classifying plant communities goes to the root of the matter; and the point of view which Gradmann here states with lucidity and vigour is precisely the point of view which the British school of plant geographers has definitely adopted.

On bare slopes, where the soil is very dry, shallow, and calcareous, the dominant grass (*Festuca ovina*) tends to become much less abundant; and small plant societies, dominated by such species as *Brachypodium gracile*, *Lotus corniculatus*, and *Thymus Serpyllum* become numerous and abundant.

The vegetation of the marshy places of the limestone slopes is described in Chapter VI.

Mixed Calcareous Grassland

On ascending a steep slope of calcareous pasture in one of the limestone dales, it is found that, at an elevation of about 1000 feet (305 m.), it opens out on to a plateau of upland cultivation. From this plateau, rounded hill-summits rise, the highest of which, on Bradwell moor, reaches an altitude of 1550 feet (472 m.). On the lower portions of the limestone plateau, calcareous pasture may occur; but this is here frequently fenced by characteristic white walls of loose blocks of limestone. The walls indicate that some attempt has been made to reclaim the land, and that regular grazing takes place. Where the land has been ploughed at least once, and cultivation continued, either permanent pasture or arable land still occurs (see Chapter VIII); but if the land has never been ploughed or if it has lapsed from cultivation, a type of grassland occurs which is related to the primitive calcareous grassland. Sheep, cattle, and horses may be frequently grazed over it; and thus those plants of the calcareous grassland which cannot endure a high nitrogen content of the soil die off, while the rest remain. A few other plants which tolerate the manuring of the soil invade the pasture; and thus a type of grassland occurs which is, in a general way, related to the "Fettwiesen" of the Swiss plant geographers (cf. Brockmann, 1907: 332; Rübel, 1911: 143).

Analogous vegetation occurs, of course, on the grasslands of the sandstones and shales. However, the details of British grassland associations, or groups of associations, have not yet been fully investigated.

On the limestone plateaux, such tracts of grassland are frequently characterized by the great abundance of the mountain pansy (*Viola lutea* and *V. lutea* var. *amoena*), which, in early summer when the plant flowers with great exuberance, gives rise to a distinct facies or aspect.

Transitional Calcareous Grassland

At still higher altitudes (about 1100 feet = 335 m.), a type of pasture occurs which is transitional between calcareous grassland and siliceous grassland. A certain number of heath-loving or humus-loving species, such as *Luzula multiflora*, *Potentilla erecta*, *Polygala serpyllacea*, and even *Ulex Gallii*, and also a certain number of lime-loving species, particularly *Poterium Sanguisorba*, may occur; and, under these circumstances, such grassland may be regarded as transitional between calcareous and siliceous grassland. Such grassland occurs also on the Carboniferous Limestone in Yorkshire. Analysis shows that the superficial soil of such localities is comparatively poor in lime, even when the subsoil consists of limestone, and, as stated in the introductory chapter, even when there are no glacial or other foreign deposits. Doubtless the soil, in the course of ages, has had much of its lime carried away in solution. This type of grassland is, on the vegetation maps, given the same colour as that used for siliceous grassland.

Calcareous Heath

Where the lime-content of the superficial layer of soil is still lower, such dwarf-shrubs or under-shrubs as the heather (*Calluna vulgaris*) and the bilberry (*Vaccinium Myrtillus*) may occur; and thus transitions occur between calcareous grassland and heath. This transitional association may be termed a calcareous heath. The association occurs on the Carboniferous Limestone in Somerset (Moss, 1907 *a*: 46), usually at altitudes above 600 feet (183 m.), below which altitude, typical calcareous grassland is there the rule. In the west of Ireland, calcareous heath is widespread on an extensive lowland plain of Carboniferous Limestone in co. Clare (see *The New Phytologist*, 1908: 259). The calcareous heaths of the present district are rather meagrely developed; but examples occur at the head of Monk's dale north of Miller's dale, and at the east of Longstone Edge north of Longstone. Other examples occur south and south-east of Buxton. There is rather an important difference between the limestone heaths of Somerset and those of the Peak District: those of Somerset usually occur below 850 feet (259 m.) and those of Derbyshire above 1000 feet (305 m.). Correlated with this altitudinal difference, there is a considerable difference in the floristic composition of the two associations. Whereas the limestone heaths of Somerset are characterized by only some half-dozen heath-loving species (albeit these are often very numerous as regards individuals) and a very large number of lime-loving species, the limestone heaths of Derbyshire are characterized by a small number of lime-loving species (which, however, are abundant) and a large number of humus-loving species. A Derbyshire botanist may obtain a rough idea of a Somerset calcareous heath if he imagines his grassy dale-slopes to possess all the numerous lime-loving species which actually occur there, and to possess, in addition, numerous and well-grown plants of ling or heather (*Calluna vulgaris*), heath (*Erica cinerea*), gorse (*Ulex* spp.), and bracken (*Pteris aquilina*). It is a factor of some importance also that the calcareous heath of Somerset occurs on slopes which are much less steep than those of the limestone dales of the Peak District of Derbyshire.

Graebner (1901) has also stated that *Calluna* may occur on calcareous soils, not incidentally but in quantity. A mixture of calcicole and calcifuge species on calcareous soils has been described by Géze (1908: 463—4), who also states that the lime is frequently leached out of the superficial layers of soil.

It was suggested in the previous chapter that the greater percentage of lime on the soil of the steep calcareous slopes is probably due to the upper layers in such places being in process of slow but continuous denudation; and hence the superficial soil is constantly changing, and new and more highly calcareous layers brought into use for the plants. On the other hand, the surface soils on the flatter plateaux are not washed away; and hence they receive no replenishment of new soil from the subjacent rock: consequently, as leaching continues, the percentage of lime in the less steep localities becomes more and more reduced as time passes. If this reasoning be correct, it follows that calcareous heaths should be more characteristic of flatter and exposed situations than of steep hill-slopes; and this is actually the case.

It is most interesting to note that the humus-loving plants of the calcareous heath, such as the heather (*Calluna vulgaris*), are shallow-rooted plants, and that the lime-loving species, such as the burnet (*Poterium Sanguisorba*), are deep-rooted plants. Thus the roots of the lime-loving species are able to reach the lower layers of the soil where the lime-content remains high; and the roots of the humus-loving species perform their work in the upper layers where the lime-content is low and the humus-content high. The calcareous heath is therefore a complementary plant community (cf. Woodhead, 1906: 345), where species of antagonistic requirements live side by side in virtue of their roots occupying different levels in the soil.

The leaching of lime from calcareous soils has, of course, long been known; and it is to be expected in districts like the Pennines, the Mendips, and the west of Ireland, where the rainfall is high. The importance of the process in ecological plant geography is that by this means a soil may in time become so changed in character as to support a totally different group of plant associations from those which first occupied it. By this process of leaching, it is conceivable that a particular tract of calcareous pasture may ultimately disappear from a given spot and be replaced by siliceous grassland or even by heath or moor; and similarly it is possible that an ash wood may in time be superseded by an oak wood. Such a process is, in its general effects, comparable with the changes which occur in the conversion of a "Niedermoor" ("Flachmoor," in part) or fen characterised by an alkaline peat into a "Hochmoor," or true moorland, characterized by an acidic peat.

A single plant formation is, within a district of uniform climate, marked by a generally uniform type of soil. When, by any means, the soil becomes radically changed, then a new plant formation has also been called into being on the site of the former one. When, for example, a sheet of open water becomes filled up with silt and peat, the aquatic formation has succumbed and a fen or a peat moor has taken its place. There are, then, not only *intermediate associations* in any single plant formation, but also *passage associations* leading from one formation to another. The limestone heath is such a passage association. Geologists have long termed certain strata between two geological formations transitional or passage beds; and it is to be expected from the nature of the case, that similar transitional tracts of vegetation should connect certain allied plant formations. Doubtless some difference of opinion may arise as to which of two plant formations a particular passage association should be referred; but such a matter is not really one of fundamental importance.

Some of the bare or almost bare limestone rocks at altitudes approaching 1500 feet (457 m.) also furnish an interesting mixture of lime-loving and humus-loving plants. For example, the following mixture of lime-loving and humus-loving species was noted at Thirkelow rocks, south of Buxton:—

Asplenium viride
Poterium Sanguisorba
Sedum acre
Vaccinium Myrtillus
Calluna vulgaris
Thymus Serpyllum (agg.)
Galium saxatile
Deschampsia flexuosa
Festuca ovina
Nardus stricta
Luzula erecta

In the case of the heather and the bilberry, it was found that, whilst some of the roots of the plants were closely appressed to the bare limestone rock, other roots of the same plant were embedded in black humus formed chiefly of decaying lichens and mosses. It is a matter for experiment whether seedlings of these plants will germinate and arrive at maturity if grown in a calcareous soil destitute of humus.

The following list of species are illustrative of the limestone heaths of north Derbyshire:—

Lime-loving, on the whole	Humus-loving, on the whole	Ubiquitous
Arenaria verna	Pteris aquilina	Botrychium Lunaria
A. serpyllifolia	Potentilla erecta	Ranunculus bulbosus
Thlaspi alpestre	P. procumbens	Lotus corniculatus
Arabis hirsuta	Ulex Gallii	Linum catharticum
Poterium Sanguisorba	U. europaeus	Viola lutea
Anthyllus Vulneraria	Lathyrus montanus	Conopodium majus
Helianthemum Chamaecistus	Polygala depressa	Gentiana Amarella
Sedum acre	Galium saxatile	Veronica officinalis
Gentiana baltica	Erica cinerea	Plantago lanceolata
Thymus Serpyllum	Calluna vulgaris	Crepis virens
Plantago media	Vaccinium Myrtillus	Hypochaeris radicata
Galium verum	V. Vitis-idaea	Leontodon hispidus
G. sylvestre *var.* nitidulum	Deschampsia flexuosa	Hieracium Pilosella
Carlina vulgaris	Nardus stricta	Anthoxanthum odoratum
Carduus nutans	Juncus squarrosus	Agrostis vulgaris
Avena pubescens	J. effusus *forma* compactus	Briza media
Koeleria cristata (agg.)	Luzula multiflora *forma* congesta	Festuca ovina
		Carex flacca
		Luzula campestris

Pseudo-Calcareous Heaths

The true calcareous heath should not be confused with the alternation of lime-loving and humus-loving species where this alternation is due to the occurrence of non-calcareous detritus or drift in a chalky or limestone district. Such a tract of vegetation is made up of the mixture of small societies or associations belonging to two or more plant formations, just as the terrestrial vegetation of small islands dotted about a lake differs from the aquatic vegetation in the lake itself. The true calcareous heath occurs on soil where there is no drift or foreign soil of any kind; and the transitional nature of the vegetation is due to a secular and gradual change in the nature of the primitive, calcareous soil.

Again, many of the higher summits of that portion of north Derbyshire which is indicated, on the geological maps, as consisting of Carboniferous Limestone are capped by a layer of non-calcareous chert; and on this, of course, normal siliceous grassland and moorland frequently occur.

SPECIES OF THE CALCAREOUS GRASSLAND AND THE SILICEOUS GRASSLAND

The following is a fairly complete list of the plants inhabiting the two groups of associations, calcareous grassland and siliceous grassland of the southern Pennines:—

	Calcareous grassland	Siliceous grassland
Lycopodium Selago	vr	vr
L. clavatum	vr	vr
Ophioglossum vulgatum	la	la
Botrychium Lunaria	r to o	r
Pteris aquilina	r	r to ls
Lomaria spicant	—	o
Cystoperis fragilis[1]	l	—
Phegopteris Robertiana[2]	l	—
Asplenium viride[1]	vr	—
A. Trichomones[1]	la	—
A. Adiantum-nigrum[1]	l	—
Nephrodium montanum[3]	—	la
N. aristata	—	lo
Salix repens[3]	—	vr
S. aurita[3]	—	l
Quercus sessiliflora (dwarfed)	—	l
Betula pubescens (dwarfed)	—	l
Corylus Avellana	l	l
Rumex Acetosella	r	o to a
R. Acetosa	o	r to o
Dianthus deltoides	l	vr
Spergularia rubra	—	r
Arenaria verna	la	—
A. serpyllifolia	la	vr
Sagina nodosa[3]	r	vr
Stellaria Holostea	lo	lo
S. graminea	o	lo
Trollius europaeus[3]	l	vr
Ranunculus repens[3]	la	la
R. acris	o	l
R. bulbosus	a	r
Thalictrum minus	r	—
Cochlearia alpina	r	—
Thlaspi virens	r	—
"T. sylvestre"	r	—
Sisymbrium Thalianum	l	vr
Cardamine hirsuta	l	—
Draba muralis[1]	r	—
D. incana[1]	vr	—
Erophila verna	la	vr

	Calcareous grassland	Siliceous grassland
E. praecox	la	—
"E. inflata"	vr	—
Arabis hirsuta	o	—
Hutchinsea petraea	l	—
Sedum acre	a	—
*S. album (agg.)	l	—
S. Telephium[1]	r	—
"Saxifraga hirta"	r	—
S. hypnoides	la	—
S. granulata	la	r
S. tridactylites	la	—
Parnassia palustris[3]	l	vr
Spiraea Ulmaria	l	l
S. Filipendula	r	—
Crataegus Oxyacantha (dwarfed)	la	lo
Rubus spp.	r to o	r to o
Rosa spp.	r to o	r to o
Potentilla sterilis	o	r to o
"P. verna"	l	—
P. erecta	lr	o to a
P. procumbens	—	lo
P. reptans	r to o	r to o
P. Anserina[3]	l	l
Geum rivale	r	r
G. rivale × urbanum (= G. intermedium)	vr	—
Alchemilla arvensis	l	—
A. vulgaris[4]	r	r to o
var. minor	o	r
var. alpestris	r	r
Agrimonia Eupatoria	r	—
A. odorata	r	—
Poterium Sanguisorba	a	—
P. officinale	l	l
Prunus spinosa	r	r
Pyrus Aucuparia (dwarfed)	vr	r
Genista tinctoria	—	r
G. anglica	—	r
Ulex Gallii	lr	la
U. europaeus	lr	la
Cytisus scoparius	—	la
Anthyllis Vulneraria	o	—
Lotus uliginosus[3]	r	la
L. corniculatus	a	r to a
Ononis repens	?	r
O. spinosa	r	—
Medicago lupulina	o	l
Trifolium medium	lo	lo
T. pratense[4]	lo	lo
"T. striatum"	r	—
T. repens[4]	r to o	r to o

	Calcareous grassland	Siliceous grassland
T. dubium	r	r
T. filiforme	vr	—
Hippocrepis comosa	r	—
Vicia sepium	r	r
V. angustifolia	vr	r
Lathyrus montana	r	r to o
L. pratensis	o	l
Geranium molle [4]	l	l
G. pusillum [4]	vr	vr
G. dissectum [4]	l	l
G. lucidum [1]	r to a	—
G. Robertianum	o	lo
G. sanguineum [1]	r	—
Oxalis Acetosella	r	r
Linum catharticum	a	r to a
Polygala vulgaris	o	r
"P. oxyptera"	r	r
P. serpyllacea	—	o
Empetrum nigrum	—	l
Ilex Aquifolium	—	r
Hypericum pulchrum	r	r to o
H. humifusum	—	r
H. montanum	r	—
H. hirsutum	lo	—
H. quadratum [3]	l	l
Helianthemum Chamaecistus	a	—
Viola hirta	a	—
V. Riviniana	o	r to o
V. lutea	ls	r, la
var. amoena	r to o	vr
Pimpinella magna	r	—
P. Saxifraga	o	r to o
Conopodium majus	o	r to o
Heracleum Sphondylium	o	r to o
Daucus Carota	r	—
Calluna vulgaris	vr	r to a
Erica cinerea	—	r to a
E. Tetralix [3]	—	l
Vaccinium Myrtillus	vr	r to a
V. Vitis-idaea	vr	la
Primula veris	a	vr
Fraxinus excelsior (dwarfed)	r to o	vr
Centaurion umbellatum (=Erythraea Centaurium)	r	r
Gentiana Amarella	o to a	la
G. baltica	vr	vr
Myosotis collina	r	—
M. versicolor	r	r
Ajuga reptans [3]	o	r to o
Teucrium Scorodonia	l	la

	Calcareous pasture	Siliceous pasture
Scutellaria minor[3]	—	r
Nepeta hederacea	l	l
Prunella vulgaris	o	r to o
Stachys Betonica	o	r to o
Satureia Acinos	vr	—
S. Clinopodium	r	—
Origanum vulgare	o to a	—
Thymus Serpyllum (agg.)	a	r
Verbascum Thapsus	r	—
Linaria vulgaris	l	l
Veronica didyma[4]	l	l
V. serpyllifolia[4]	l	l
V. Chamaedrys	o to a	r to o
Digitalis purpurea	—	r to o
Melampyrum pratense (agg.)	—	l
Euphrasia officinalis (agg.)	a	r to la
Bartsia Odontites[3]	la	vr
Pedicularis sylvatica[3]	—	l
Rhinanthus Crista-Galli (agg.)	la	l
Plantago media	o	—
P. lanceolata	a	r to o
Sherardia arvensis[4]	l	—
Asperula cynanchica	r	—
Galium verum	a	r
G. Cruciata	la	—
G. saxatile	r	o to a
G. sylvestre[1]	la	—
var. nitidulum[1]	la	—
Lonicera Periclymenum	r	r
Valerianella carinata[1]	r	—
V. olitoria[4]	r	—
Scabiosa Succisa	r	lo
S. Columbaria	o	—
S. arvensis	r to o	r
Campanula glomerata	r	—
C. rotundifolia	r to o	o to a
Wahlenbergia hederacea[3]	—	vr
Jasione montana	—	r to o
Solidago Virgaurea	la	la
Bellis perennis	o	r to o
Antennaria dioica	vr	vr
Gnaphalium sylvaticum	—	r
Pulicaria dysenterica	l	—
Inula squarrosa	r	—
Achillaea Ptarmica[3]	—	l
A. Millefolium	o	r to o
Chrysanthemum Leucanthemum	o to a	lo
Senecio Jacobaea[4]	r to o	r to o
S. erucifolius	r	—
Carlina vulgaris	o	vr
Carduus nutans	r to o	—

	Calcareous pasture	Siliceous pasture
Cnicus eriophorus	l	—
C. palustris	o	l
Centaurea nigra	o	r to o
C. Scabiosa[1]	o	—
Picris hieracioides	r	—
Crepis virens	a	r to o
Leontodon hirtus	a	r to o
L. hispidus	a	r to o
L. autumnalis	la	r to o
Hypochaeris radicata	o	r to o
Taraxacum vulgare (agg.)	r to o	r
T. palustre	—	r
T. laevigatum	r	?
"T. corniculatum"	r	—
Hieracium Pilosella	a	r to o
"H. cymbifolium"	r	—
H. sylvaticum	la	—
"H. holophyllum"	r	—
H. vulgatum	la	r to o
H. sciaphilum	l	l
H. rigidum	r	r
H. boreale	r	r to la
H. umbellatum	r	vr
Anthoxanthum odoratum	la	r to o
Phleum pratense[4]	r to o	r to o
Agrostis alba[3]	la	la
A. tenuis (=A. vulgaris)	a	la
A. canina[3]	—	l
Holcus lanatus[3]	o	o
H. mollis	—	l
Aira praecox	—	l
A. caryophyllea	lo	—
Deschampsia flexuosa	—	s
D. caespitosa[3]	la	la
Trisetum flavescens	o	r
Avena pratensis	r	—
A. pubescens	r	—
Arrhenatherum elatius	a	l
Triodea decumbens	r	lo
Cynosurus cristatus	la	la
Molinia caerulea	—	l
Koeleria cristata (agg.)	o	—
Briza media	o to a	lo
Dactylis glomerata[4]	lo	l
Poa annua[4]	r	r
P. pratensis[4]	l	l
Festuca ovina	a to d	r to la
F. rubra	r	r
F. elatior (agg.)	r	r

	Calcareous grassland	Siliceous grassland
"Bromus erectus"	r	—
B. mollis[4]	l	l
Brachypodium sylvaticum	a	l
B. pinnatum	vr	—
Nardus stricta	—	s
Scirpus compressus[3]	vr	—
Carex caryophyllea	la	l
C. ovalis	l	l
C. disticha[3]	vr	—
C. flacca	a	lo
C. Goodenoughii[3]	—	la
C. ornithopoda[1]	r	—
C. pilulifera	r	o
C. binervis	—	o
C. hirta[3]	l	r
Juncus inflexus (=J. glaucus)	la	lr
J. effusus[3]	r	la
J. squarrosus[3]	—	la
Luzula vernalis	r to o	r to o
L. campestris	a	la
L. multiflora	—	o
forma congesta	—	r to o
*Allium vineale	vr	—
A. oleraceum	vr	—
Ophrys apifera	vr	—
"O. muscifera"	vr	—
Orchis morio	r	—
O. mascula	la	r
O. maculata	lo	?
O. ericetorum	—	r
O. ustulata	r	—
O. pyramidalis	vr	—
Habenaria viridis	r	vr
H. bifolia	vr	vr
H. conopsea	r	r
"H. albida"	—	vr
Listera ovata	r to o	lr
Spiranthes autumnalis	vr	—

[1] Chiefly on rocks. [2] Chiefly on screes. [3] In marshy places.
[4] Chiefly invaders from the permanent pasture.

Relationships of the Plant Associations of the Siliceous and the Calcareous Soils

Precisely the same line of reasoning which decides one to place the oak and birch woods in the same plant formation as the scrub and the grassland of the siliceous soils decides one also to place the ash woods and the scrub and the grassland of the calcareous soils in another single formation.

The parallel relationships of these formations and associations, as developed on the southern Pennines, may be expressed diagrammatically in the following manner:—

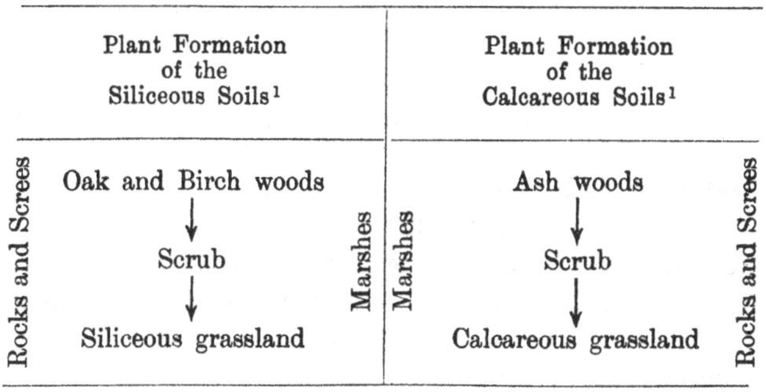

	Plant Formation of the Siliceous Soils[1]			Plant Formation of the Calcareous Soils[1]	
Rocks and Screes	Oak and Birch woods ↓ Scrub ↓ Siliceous grassland	Marshes	Marshes	Ash woods ↓ Scrub ↓ Calcareous grassland	Rocks and Screes

[1] Cf. Grisebach's (1846 : 73 and 156, and 1849 : 340—342) usage of the term "plant formation," and also Schimper's (1903 : 161).

CHAPTER V

ASSOCIATIONS OF ROCKS AND SCREES

The limestone cliffs: limestone screes. Sandstone rocks and screes. Are the plants of cliffs and screes lithophytes?

THE LIMESTONE CLIFFS

LIMESTONE cliffs are very numerous in the district, and some of the gorges, as the one known as the Winnats, near Castleton, are two hundred feet (61 m.) deep. It appears to be generally accepted among geologists that such limestone gorges represent ancient underground water-ways whose roofs have collapsed.

On the damper and more sheltered cliffs, ferns and flowering plants occur in the crevices and on the ledges. Most of these plants are members of the neighbouring plant associations, such as ash woods, scrub, and calcareous pastures. Near villages, as on the cliffs near Middleton at the foot of Middleton Dale, several alien plants have established themselves. The richness of the vegetation of the limestone cliffs varies with their dampness, the damper cliffs being rich in species, the driest ones extremely poor. The dampness or dryness of the cliffs is largely determined by the dip of the strata; and hence, in any gorge, the rocks on one side are usually richer in plants than the rocks on the opposite side. For the same reason, the vegetation of the opposite sides of a valley may vary somewhat in character. Aspect alone does not usually appear to be a fundamental differentiating factor, except in the case of species at or near their limit of distribution.

In the following list of the more characteristic plants

Copyright *W. B. Crump*

Figure 18.

Ash Wood and Limestone cliffs.
River Wye flowing between Limestone cliffs.

occurring on the limestone cliffs of Derbyshire, the species which do not occur on the sandstones or shales are preceded by the letter "L." The list is by no means an exhaustive one, as a full list would contain many of the plants of the ash woods, related scrub, and calcareous pasture. However, certain species are more partial to the cliffs than to any other kind of habitat. The following are species of this character:—

Liverworts.

Frullania Tamarisci
L. F. dilatata
L. Lejeunia calcarea
L. L. serpyllifolia
L. L. Rosettiana
L. Porella platyphylla
L. Scapania aequiloba
S. aspera
Pedinophyllum interruptum
Jungermannia riparia
L. J. turbinata
J. bantriensis
L. Metzgeria pubescens
M. furcata
L. Reboulia hemisphaerica
Riccia glauca

Mosses.

Ditrichum flexicaule
"Swartzia montana"
L. Selegeria Doniana
L. S. pusilla
L. S. acutifolia
L. "S. tristicha"
L. "S. calcarea"
Fissidens spp.
F. decipiens
F. adantioides
Grimmia apocarpa
G. pulvinata
Rhacomitrium canescens
R. lanuginosum
Tortula muralis
T. subulata
Barbula rubella
B. tophacea
L. Weissia calcarea
W. rupestris
Trichostomum crispulum
L. T. mutabile
T. tortuosum
Encalypta vulgaris
E. streptocarpa
Zygodon viridissimus
Orthotrichum anomalum
O. cupulatum
L. Funaria calcarea
F. hygrometrica
L. Bartramia Oederi
L. Webera cruda
Bryum spp.
B. capillare
Minum spp.
M. stellare
Neckera crispa
Anomodon viticulosus
Pleuropus sericeus
Campothecium lutescens
Eurynchium murale
E. tenellum
L. E. Teesdalei
E. pumilum
E. Swartzii
E. crassinervium
Amblystegium spp.
L. A. confervoides
Hypnum spp.
H. molluscum

Vascular plants.

L. Polypodium vulgare. Rare
L. Asplenium viride. Rare
L. A. Trichomones
L. A. Ruta-muraria
L. A. Adiantum-nigrum
L. Ceterach officiniarum. Rare
L. Phyllitis Scolopendrium. Rare
L. Cystopteris fragilis

L. Taxus baccata. Local
L. "Juniperus communis." Rare
Corylus Avellana
Urtica dioica
L. Parietaria officinalis. Local
Lychnis dioica
L. Silene nutans. Local
L. Thalictrum minus. Local
L. *Arabis albida. Local
L. A. hirsuta
L. Cardamine impatiens
L. Hutchinsea petraea. Local
L. Draba muralis
L. D. incana. Rare
Erophila vulgaris
L. E. praecox. Local
L. Cochlearia alpina. Local
Sisymbrium Thalianum
L. *Cheiranthus Cheiri. Local
L. *Iberis amara. Local
Cardamine flexuosa
L. C. hirsuta
L. Viola hirta
L. V. sylvestris
V. Riviniana
L. *var.* villosa
L. Sedum acre
S. anglicum. Local
L. S. Telephium
L. "Saxifraga hirta." Rare
L. S. hypnoides
L. S. tridactylites
L. *S. umbrosa. Rare
L. Pyrus Aria. Local
P. Aucuparia. Local
L. Poterium Sanguisorba
L. Geranium lucidum
L. G. sanguineum. Local
G. Robertianum
Linum catharticum
Polygala vulgaris
"P. oxypteris." Rare
L. Helianthemum Chamaecistus
L. Pimpinella magna

Fraxinus excelsior
L. *Lamium maculatum
Myosotis sylvatica
M. arvensis
var. umbrosa
L. Galium sylvestre
L. *var.* nitidulum
Valerianella olitoria
L. V. cruciata
L. *Kentranthus rubra
L. Scabiosa Columbaria
L. Centaurea Scabiosa
*Chrysanthemum Parthenium
C. Leucanthemum
*Doronicum paraloides
Lactuca muralis
Taraxacum officinale
T. erythrospermum
"T. laevigatum"
Hieracium Pilosella
H. brittanicum
"H. rivale." Rare
L. "H. cymbifolium." Rare
H. sylvaticum
L. "H. rubiginosum." Rare
L. "H. holophyllum." Rare
H. vulgatum
"H. diaphanoides"
H. sciaphilum
H. rigidum
"H. prenanthoides"
H. boreale. Rare
H. umbellatum. Local

Arrhenatherum avenaceum
Brachypodium gracile

L.	Melica nutans. Local	L.	C. ornithopoda. Local
	M. uniflora		C. pallescens. Local
	Festuca ovina		Tamus communis
	Poa nemoralis. Local	L.	Convallaria majalis. Local
	Carex pulicaris. Local		

Limestone pavements, which are so characteristic of the limestone plateau of the mid-Pennines (see Smith and Rankin, 1903: 167) and of the lowland limestone plain of Co. Clare (see *New Phytologist*, 1908: 258) scarcely occur in the Peak District of Derbyshire.

Ostenfeld (1908: 972), in his account of the vegetation of the cliffs of the Faeröes, states that the water-content of the soil, "before all others is the factor which has the greatest influence, and is the first and most important condition in differentiating between plant associations with the same geographical and topographical position": this remark is doubtless true when the rocks and soils in question are of a similar chemical composition; but such a classification of the plant associations of a district which, like the Peak District, consists on the one hand of sandstone rocks and siliceous soils and of limestone rocks and highly calcareous soils on the other, would give a very queer and a most unnatural result. Water-content alone fails to supply a primary differentiating factor of the plant associations in a district like this where sandstone rocks are sharply contrasted with limestone rocks. The only primary factor giving a natural classification of the plant associations of the terrestrial soils of this district is one based on the presence as contrasted with the comparative absence of lime in the soil. Secondarily, or when applied either to the siliceous or to the calcareous soils alone, water-content becomes a decisive ecological factor; but even this is complicated by the acidic humus-content of many of the siliceous soils.

Limestone Screes

The screes consist of angular pieces of rock, a few inches in diameter on the average, which have fallen from the disintegrating cliffs above. Such stretches of weathered *débris* are of common occurrence on the slopes of the hills in the limestone area. The screes of this district, however, are not specially well developed; and in no cases are they difficult or dangerous to traverse. The vegetation of the limestone screes of Somerset has been previously described (Moss, 1907 *a*: 49); and, of the species of plants there mentioned, all except the Welsh poppy (*Meconopsis cambrica*) occur in Derbyshire, though in this locality the scaly fern (*Ceterach officinarum*), the harts-tongue (*Scolopendrium vulgare*), the yew (*Taxus baccata*), and the whitebeam (*Pyrus Aria*) are much rarer than in Somerset.

Here, the screes are never of great depth; and very often plants, whose aërial parts appear above the loose talus, are rooted in the soil below. Such soil does not differ materially from the rest of the soil of the limestone slopes, but, being covered by stones, evaporation is less intense. Thus, a few moisture-loving species, such as *Allium ursinum*, *Geranium Robertianum*, *Mercurialis perennis*, *Scrophularia nodosa*, and *Valeriana sambucifolia*, apparently occur on the older screes. Closer examination, however, proves that all these plants are really rooted in the soil below the screes. There are, in fact, no true "lithophytes" on the screes of Derbyshire or Somerset, except perhaps the lichens and some of the mosses that grow on the bare rocks and stones themselves.

The screes, however, are interesting as they furnish examples of open associations. Doubtless, in most cases, woodland or scrub or grassland characterized the hill-slopes which are now covered by the screes before the latter fell away from the rocky escarpment above. The *débris* would destroy the original plant associations; and the new surface would thus afford a suitable habitat for the invasion of plants from the neighbouring associations. Newly formed screes, since they have very little vegetation, may be regarded as edaphic deserts. In fact, probably all open plant associations, in all non-arctic or non-alpine districts, which have a mean annual rainfall of fifteen inches (28 cm.) or more, may be so regarded. Only those plants which have long subaerial organs are able to colonize the newer screes. Where the screes are continually, though perhaps slowly, accumulating, the plant associations remain in an open condition. On such new screes, the following plants have been observed thinly scattered about:—

Phegopteris Robertiana
  (= Polypodium calcareum)
Arrhenatherum avenaceum
Brachypodium gracile
Corylus Avellana (dwarfed)
Geranium Robertianum
Teucrium Scorodonia
Fraxinus excelsior

On the older screes, the plant associations tend to become more and more closed; and it is well known (cf. Warming, 1909 : 246) that screes often show a developmental history. In this district, as in Somerset, three types of plant succession may be recognised as characteristic of the screes. The most frequent case is the succession which terminates in calcareous grassland. A not uncommon succession terminates in an ash wood, and intermediate stages of this succession are well shown on screes in Haydale, east of Cressbrook Dale. The least frequent succession leads on to a kind of limestone heath, as at the head of Monksdale, north of Miller's dale, where *Calluna vulgaris* occurs side by side with lime-loving plants: Smith and Rankin (1903: 167) mentioned that a similar kind of vegetation is seen on some of the limestone screes of the mid-Pennines.

The following list was compiled from older screes adjoining an ash wood:—

Phegopteris Robertiana
Cystopteris fragilis
Polypodium vulgare
Corylus Avellana
Urtica dioica
Thalictrum minus
Sedum acre
Saxifraga hypnoides
Rubus saxatilis
Crataegus Oxyacantha
Geranium lucidum
G. sanguineum
G. Robertianum
Oxalis Acetosella
Mercurialis perennis
Cornus sanguinea
Scrophularia nodosa
Teucrium Scorodonia
Galium sylvestre
Sambucus nigra
Campanula rotundifolia
Valeriana sambucifolia
Valerianella olitoria
V. carinata
Senecio Jacobaea
Solidago Virgaurea
Picris hieracioides
Arrhenatherum avenaceum
Brachypodium sylvaticum
Melica nutans
Convallaria majalis
Allium ursinum

On higher mountains than occur in Derbyshire, screes are developed to a correspondingly great extent: and the stones composing the screes may then be many yards in diameter. The vegetation of such block-screes is usually extremely scanty, as the large size of the boulders prevents so much light from reaching the soil below that seedling plants are unable to reach maturity. Such tracts are well known in the Alps, and have been described by the Swiss plant geographers under the name of Geröllflur or Geröllflora or Schuttflora (cf. Öttli, 1905: 18). Brockmann (1907 : 290—1) subdivides his "formations group" of the Geröllflora into plant communities (*a*) on siliceous rocks, and (*b*) on calcareous rocks, and gives lists of plants for each subdivision. It may well be that in districts like the Alps, where the great differences in altitude produce very marked differences in the vegetation at different heights, the vegetation of the Alpine boulder-strewn ground belongs to a different plant formation from other parts of the mountain slope; but in this district, where the differences in altitude on the limestone hill-slopes are comparatively slight, and where the depth of the *débris* of stones is rather insignificant, the plant communities seen on the screes can scarcely be separated from those on the other parts of the hill sides (cf. figure 21).

SANDSTONE ROCKS AND SCREES

Screes and boulder-strewn slopes also occur to some extent on the siliceous slopes below escarpments of the Carboniferous gritstones; but here also the flora partakes of the same general composition as that of the associations in close propinquity. For example, the sandstone screes in the moorland area are characterized by such plants as *Calluna vulgaris, Vaccinium Myrtillus, V. Vitis-idaea, Arctostaphylos Uva-ursi*, and *Deschampsia flexuosa*, which are rooted not on the bare sandstone rocks but in the soil in which the boulders are embedded, or in the clefts of the rocks, or on soil which has accumulated on the projecting ledges.

Most of the sandstone rocks and screes of the district occur in the moorland area. Occasionally they occur in woods, and only rarely in the grassland or cultivated areas. As in the case of the limestone rocks and screes, the plants present belong for the most part to the adjoining associations. For example, on the numerous "edges" or sandstone escarpments of the moorland area, humus collects in the rocky crevices, and on the rocky ledges; and here moorland plants prevail (see figure 19), particularly the bilberry (*Vaccinium Myrtillus*). The same remark applies to the sandstone screes of the moorland area, though, as these receive a considerable amount of

Copyright *A. Wilson*

Figure 19.

Rocks of Millstone Grit.

Bilberry (*Vaccinium Myrtillus*) on the ledges; and a rock-moss (*Andreaea Rothii*) on the face of the cliff at the extreme left.

shelter from the cliffs above them, they have a rather richer flora (see Chapter VII).

Although a fairly long list of cellular flowerless plants, which occur on the faces of the sandstone rocks, is given below, most of the species are very rare and local; and it is scarcely possible to single out any vascular plants which, in this district, exhibit any pronounced partiality for living on the sandstone rocks. Many of the cellular cryptogams are very susceptible to the action of smoke (Wilson, 1900); and, as the southern Pennines are situated between two great manufacturing districts, it is highly probable that many of the mosses and lichens characteristic of bare rocks are even rarer now than they were a century ago. The following silicolous and saxicolous cellular plants have been recorded (Linton, 1903; Crossland, 1904; *etc.*) for the sandstone rocks of the southern Pennines: the species which are confined to such rocks are preceded by the letter "S":—

S. Andreaea Rothii
S. A. crassinerva
S. A. alpina
S. A. petrophylla
S. Tetraphis Browniana
S. Swartzia montana
S. Dicranum fuscescens
Grimmia apocarpa
G. pulvinata
G. trichophylla
G. Doniana
S. Rhacomitrium fasciculare
S. R. heterostichum
R. lanuginosum
R. canescens
Phytomitrium polyphyllum
S. Campylosteleum saxicola
S. Hedwigia ciliata
Tortula muralis
Eurhynchium murale
Leconora, ? sp.
Lecidea, ? sp.
S. Parmelia saxatilis
S. Pertusaria dealbata

Crampton has recently described the vegetation of the screes of Caithness. The plants of these screes are chiefly humus-loving species, such as frequently occur on the sandstone screes of the Pennines; and there would appear to be little justification for giving the vegetation in question the rank either of "formation" or even "subformation" (Crampton, 1911: 26 and 43). This will perhaps best be seen by quoting all the species mentioned by Crampton. Those which do not occur on the Pennines are indicated by a †:—

Sphagnum spp.
Rhacomitrium lanuginosum
Hylocomium spp.
Hypnum Schreberi
†Silene amoena
†Alchemilla alpina

Rubus Chamaemorus
Empetrum nigrum
Erica cinerea
Calluna vulgaris
†Azalea procumbens
Vaccinium uliginosum
V. Myrtillus
V. Vitis-idaea
Melampyrum pratense
Galium saxatile
Anthoxanthum odoratum
Agrostis tenuis
Deschampsia flexuosa
Festuca ovina
Luzula sylvatica
L. campestris

ARE THE PLANTS OF THE CLIFFS AND SCREES LITHOPHYTES?

The precipitous faces of the cliffs are tenanted by many species of Algae, lichens, liverworts, and mosses; and some of these may be regarded as "lithophytes." Many of the plants, however, even the lower cryptogams, are not rooted on the bare rock itself, but in the loose soil which accumulates, to a slight extent, on the surface of the slight irregularities of the face of the rock, even when this is nearly vertical.

Warming (1909: 238 and 240) uses the term "lithophytes" in a double sense. Section VIII of Warming's treatise is headed "lithophytes"; and these are subdivided into (1) "lithophytes" and (2) "chasmophytes." Warming states that this subdivision is in accordance with the suggestion made by Schimper (1903—4: 178) who wrote:—"The vegetation on the surface of rocks or stones may be termed that of *lithophytes*. Crevices in rocks, in which more finely grained components and more water accumulate than on the surface, produce a somewhat more copious vegetation, that of *chasmophytes*." Warming (1909: 240) also cites Öttli who defines as rock-plants or petrophytes "all those plants, growing on sides of rocks or blocks of detached stone, which are able, as the first of their kind, to colonize the rock permanently, and which display in distribution or structure a more or less pronounced dependence upon rock as a substratum. Within this definition are included both lithophytes and chasmophytes." Öttli (*loc. cit.*) maintains that "it is not a natural scheme to co-ordinate both lithophytes and chasmophytes; and he suggests the following scheme:—

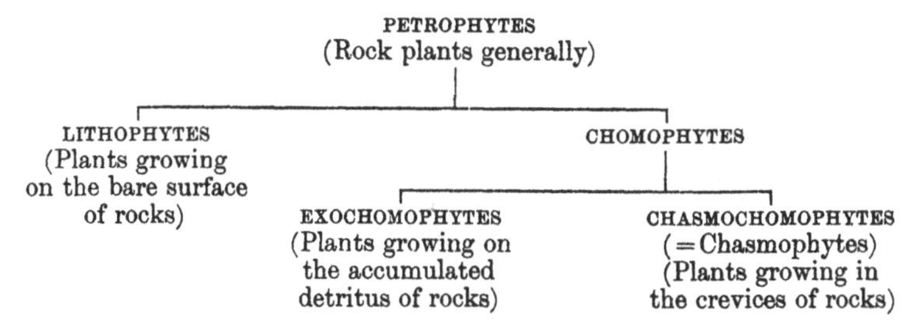

Of lithophytes (using the term in the strict sense) there are probably only certain Algae, lichens, liverworts, and mosses, *i.e.*, plants which are able to absorb atmospheric moisture by means of their general superficial tissues. It is doubtful if those plants on rocks, even including Algae and lichens, which absorb moisture by means of roots or root-like structures, should be placed in a single plant formation. At all events, no such "formation" is recognized in the present book, although a subdivision of rock plants, like that of Öttli's, is very useful from many points of view. As regards this district, it seems sufficient to regard the vegetation of the limestone rocks and screes as belonging to the plant formation of calcareous soils, and the vegetation of the sandstone rocks and screes as belonging to the plant formation of siliceous soils except where the plants occur on the acidic humus of the rock ledges (see figure 19), when the vegetation would appear to be best placed in the moorland formation.

CHAPTER VI

MARSH AND AQUATIC ASSOCIATIONS

General distribution of the marsh (or swamp) and aquatic associations. Non-calcareous waters. Swamps on the sandstones and shales. Calcareous waters. Swamps on the limestone. Ruderal marsh species. Reed swamps. The vegetation of quickly flowing streams. Alien aquatic plants. The relation of mineral salts to the flora and vegetation.

General Distribution of the Marsh (or Swamp) and Aquatic Associations

In the Peak District, as on the Pennines generally, aquatic and marsh associations, and more especially the former, are very meagrely represented. The Pennines, consisting of fissured rocks like the Carboniferous sandstones, and porous rocks like the Carboniferous limestone, have no natural lakelets or tarns such as occur on the older Silurian and Cambrian rocks. Again, in the Peak District, there are no extensive alluvial flats; and it is in such situations that aquatic and marsh associations attain a maximum degree of development. From such lowland alluvial deposits, some of the more cosmopolitan aquatic and marsh plants spread up the streams, where they form narrow, fringing associations which are too small, however, to be marked on vegetation maps except such as are constructed on a very large scale. Hence aquatic associations and reed swamps are poorly developed and only of local occurrence on the Pennines.

In this district, marshes or swamps are characteristic of those spots on the hill-slopes where springs issue, and of the immediate banks of the streams where these banks happen to be flat. Reed swamps are very local; and even when they do occur, they are very small and not very typical. The streams themselves are tenanted by numerous characteristic mosses, liverworts, and Algae, but by few characteristic aquatic flowering plants.

In a sense, some of the moorland associations described in the next chapter are aquatic, and were so classified by Warming (1895) in his earlier work on plant communities. The cotton-grass moor (see page 183), for example, possesses a soil which, during very considerable periods of most years, is supersaturated with moisture; and many members of the cotton-grass association possess such aquatic structural adaptations as aëration channels in their stems and leaves. It is now, however, very generally held (see Schimper, 1903—4; and Warming, 1909) that it is the physiological and not the physical wetness of the soil that determines whether or not plants are really hydrophilous; and, as peaty soils are now frequently regarded as being physiologically dry, moorland plants are not now usually placed among hydrophytes. On the other hand, Clements (1907: 170) maintains that the aquatic adaptations found in many moorland species are normal, that the xerophytic adaptations which they possess were acquired during some past period when the plants in question inhabited dry habitats, and that the xerophytic structures have persisted. It is indeed necessary to remember that very little experimental investigation has yet been performed on the physiological water-contents of soils, most ecologists and plant geographers being apparently content with general impressions and general statements.

As Schimper has pointed out (1904: 781), "every classification of the aquatic flora commences with the separation of salt-water forms from fresh-water forms." All the aquatic plants of this district belong to the latter class; and a classification of fresh-water aquatics may be based on the richness or poverty of the water in soluble mineral salts. From this point of view, the aquatic species of the calcareous streams of the district may be placed in one association, and those of the non-calcareous streams in another.

The rivers of lowland districts are probably always rich or fairly rich in soluble mineral salts, as the streams have in their earlier courses dissolved much material from the rocks through which they have cut their way; but this statement does not apply to the small streams of non-calcareous hill-slopes. In the Peak District, it is easy to distinguish the vegetation of the streams which flow over the non-calcareous or siliceous soils and which have a very low mineral content from that of the streams which flow over the calcareous soils and which have a high mineral content.

NON-CALCAREOUS WATERS

The late Mr Ackroyd (1899: 120), formerly the Borough Analyst of Halifax, has published figures giving the composition of the waters of a reservoir supplying that town; and this analysis is useful as it is typical of the whole of the non-calcareous waters of the sandstones and shales of the Pennines, including the Peak District. "The matters dissolved consist of mere traces of inorganic bodies, and a small amount of peaty acid....These waters are very soft, ranging from two to three Clark's degrees[1]; and the hardness is of a permanent character[2], *i.e.*, it is not appreciably lessened on boiling." The following is Mr Ackroyd's full analysis:—

	Grains per gallon	Grams[3] per litre
Total solids in solution	5·25	0·075
Total solids in suspension	nil	nil
Chlorine, calculated as $NaCl$	1·3	0·019
Iron, calculated as Fe_2O_3	0·03	0·004
Sulphate of lime ($CaSO_4$)	3·90	0·006
Free and albuminoid ammonia (NH_3)	nil	nil
Acidity, calculated to its equivalent of sulphuric acid (H_2SO_4)	0·16	
Hardness = 3 degrees (Clark's)[1]		

SWAMPS (OR MARSHES) ON THE SANDSTONES AND SHALES

The larger streams of the district have their sources on the peat moors, the smaller ones on the hill slopes. In the former case, the sources of the streams occur either in the

[1] A Clark's degree is one grain of carbonate of lime ($CaCO_3$) per gallon, or its equivalent of other lime (calcium) compound.

[2] Permanent hardness of water is due to sulphate of lime ($CaSO_4$), temporary hardness to carbonate of lime ($CaCO_3$).

[3] 15·43 grains = 1 gram; 0·22 gallons = 1 litre.

 W. B. Crump

Figure 20.

A Dripping Shaly Bank.

Lady Fern (*Athyrium Filix-foemina*), Foxglove (*Digitalis purpurea*), Wall-lettuce (*Lactuca muralis*), Golden Saxifrage (*Chrysosplenium oppositifolium*), mosses, Hepatics, Algae, etc.

heather association or in the cotton-grass association: in the latter case, they occur in associations either of siliceous grassland, scrub, oak or birch wood. In all cases, however, the source is marked by a swamp; and these swamps are frequently, but by no means invariably, dominated by the common rush (*Juncus effusus*).

As might be inferred from the diverse surroundings of these springs, there is a certain amount of diversity in their flora. This difference, however, is less than might have been anticipated, as the springs are all characterized by some trickle of water implying a fair degree of aëration, and all are characterized by a low mineral-content, as is sufficiently indicated in the analyses just given.

The poorest flora is found in the swamps surrounding the springs which occur on the peat moors. It is sometimes stated that the larger rushes do not occur on peat; but Pethybridge and Praeger (1905: 171) find an association *Juncus effusus* on peaty soils, and this is also the case on the Pennines; so that the statement is too sweeping. It is true that *Juncus effusus* is not a normal member of any of the plant associations on peat with stagnant water; but whenever water from a spring oozes through the peaty soil, there a community of the common rush frequently occurs. The springs change their positions from time to time, as may be seen by comparing the present springs with those marked on the old "six-inch" Ordnance maps. Probably when the peat was being originally formed, the land round the then springs would be destitute of peat; but when a spring at some later time issued from a different place, two local changes in the vegetation would be called into being. First, the swamp plants of the original spring would, after a time, die; and their places would be filled by plants from the neighbouring moorland association, and peat would thus form on the site of the old spring; and secondly the aërated waters of the new spring trickling through the peat at a different place would result in the death of many of the moorland species of that place, and in the invasion of some swamp-inhabiting species.

On the non-peaty slopes and at lower altitudes, the swamps are more numerous and much richer in species. However, the characteristic feature of all of them, whether on peat or not, is that there is some flow or trickle of water, *i.e.*, the water is well aërated. The swamp association of siliceous soils is typically dominated by the common rush (*Juncus effusus*), yet it is a different plant community from the Juncus facies of the siliceous grassland previously described (see page 108). In this district, the swamps of *Juncus effusus* are not of sufficient size to be given a special colour on the maps, although Pethybridge and Praeger (1905) were able to do this on their vegetation map of the district lying south of Dublin.

The term "Juncetum" has been used by several writers to denote a plant association whose chief constituent is a species of the genus Juncus. Such terms were first used by the Danish plant geographer Schouw (1822); but the term "Juncetum" is vague, as it does not indicate the species of Juncus which is the dominant plant. In this district, for example, there are several kinds of "Junceta." The associations at the stream sources have usually *Juncus effusus* as the chief constituent plant; but, besides this, associations or societies occur of *J. sylvaticus* (= *J. acutiflorus*), of *J. lamprocarpus*, of *J. supinus*, and of *J. squarrosus* on the sandstones, shales, or peat; of *J. glaucus* on the limestones chiefly; and of *J. bufonius* on almost any soil. In certain lowland localities, associations of *J. subnodulosus* (= *J. obtusiflorus*) occur, and, in maritime localities, of *J. maritimus*, locally of *J. acutus*, and of *J. Gerardi*. Thus the term "Juncetum" is very vague and ambiguous. To overcome the ambiguity, Cajander (1903) has adopted a modification of Schouw's plan. Cajander would add the specific name as a genitive to the term "Juncetum": thus, an association of *Juncus effusus* would be denoted by the term "Juncetum effusi[1]." This plan is capable of universal application with regard to pure associations; and hence is of great merit. Of course, such universal terms will not supersede the common or vernacular names of plant associations, just as *Bellis perennis* is still known as the daisy and just as trichlormethane is still known as chloroform. However, many associations are destitute of common names; and, in any case, the technical names possess all the advantages of the binomial names in general use among botanists and zoologists, and of the technical names in use among organic chemists (cf. Moss, 1910 *b*).

[1] An abbreviation of Juncetum Junci-effusi.

At altitudes from 1000 feet (305 m.) downwards many of the Juncus swamps are rich in species: above that altitude, the swamps become floristically poorer and poorer.

The full list of plants of the Juncus swamps is a very long one, in which respect the association resembles most other mixed or unstable or kinetic associations. In the following list the plants more characteristic of peaty localities are printed in italics:—

Equisetum limosum	l
E. palustre	l
E. sylvaticum	a
Nephrodium montanum	r to o
Athyrium Filix-foemina	r to o
var. convexum	r
Salix aurita	la
× cinerea	l
S. cinerea	a
var. aquatica	o
var. oleifolia	o
S. repens	*vr*
Betula pubescens	v
Rumex Acetosa	o
Montia fontana forma rivularis	la
M. fontana *var.* minor	*r*
Lychnis Flos-cuculi	o
Stellaria graminea	r to o
S. uliginosa	o to a
Ranunculus repens	la
R. Flammula	*o to a*
forma *tenuifolius*	*l*
R. hederaceus	vr
R. Lenormandi	*la*
Trollius europaeus	vr
Cardamine amara	l
C. pratensis	o
C. flexuosa	r to o
Drosera rotundifolia	*vr*
Chrysosplenium alternif.	l
C. oppositifolium	la
Spiraea Ulmaria	a
Geum rivale	r to la
× urbanum	
(= G. intermedium)	r
Potentilla palustris	*vr*
Lotus ulginosus	a
Empetrum nigrum	*l*
Hypericum humifusum	r to o
H. quadrangulum	o to a
Viola palustris	*la*
Epilobium obscurum	lo
E. palustre	o to a
Hydrocotyle vulgaris	*la*
Oenanthe crocata	r
Angelica sylvestris	o
Vaccinium Oxycoccus	*l*
Lysimachia nemorum	r to o
L. Nummularia	vr
L. vulgaris	vr
Menyanthes trifoliata	*vr*
Scutellaria minor	vr
Ajuga reptans	la
Myosotis repens	*o to a*
M. caespitosa	o
M. palustris	r
var. strigulosa	vr
"Scrophularia alata"	
(= S. umbrosa)	l
S. cinerea	
var. aquatica	l
Scrophularia nodosa	r to o
Veronica scutellata	*vr*
V. Beccabunga	la
"*Pedicularis palustris*"	*vr*
P. sylvatica	r to o
Pinguicula vulgaris	*r*
Galium Witheringii	a
Valeriana dioica	r
V. sambucifolia	la
Scabiosa Succisa	*la*
Wahlenbergia hederacea	*r*
Achillaea Ptarmica	o
Senecio aquaticus	o to a
Cnicus palustris	a
Crepis paludosa	l
Taraxacum palustre	r
Potamogeton polygonifolius	*l*
Triglochin palustre	r
Agrostis alba	o
A. canina *var.* mutica	o
Deschampsia caespitosa	a

Holcus lanatus	o	C. ampullacea	vr
Molinia caerulea	*l*	Juncus effusus	ls
Glyceria fluitans	r to a	*forma* compactus	ls
Scirpus setaceus	l	J. inflexus	
Eriophorum angustifolium	*l*	(=J. glaucus)	lr
Carex ovalis	r to o	J. sylvaticus	
C. echinata	a	(=J. acutiflorus)	ls
C. paniculata	*vr*	J. lamprocarpus	a
C. Goodenoughii	*l*	J. supinus	la
C. flacca	o	*J. squarrosus*	*l*
C. pallescens	vr	*Luzula multiflora*	*la*
C. panicea	o	forma *congesta*	*o*
C. pendula	vr	*Narthecium Ossifragum*	*l*
"C. strigosa"	vr	Iris Pseudacorus	lr
C. helodes		Orchis latifolia	?
(=C. laevigata)	la	O. maculata	?
C. binervis	o	O. ericetorum	o
C. fulva	r		
C. flava	la		
forma *oedocarpa*			
(=*C. flava* var. *minor*)	*la*		

Calcareous Waters

The composition of the calcareous waters may be inferred from the following analyses of the waters of the Carboniferous Limestone area of the mid-Pennines published by Mr Burrell (1900). It will be seen that the iron-content is practically the same as that of the non-calcareous waters, but that the lime-content is considerably higher. The magnesium-content is also high; and this fact has possibly some significance. It seems to be the case generally, although perhaps not so universally, that natural waters with a high lime-content have also a high general mineral-content, and conversely that waters with a low lime-content have also a low general mineral-content; and it is by no means improbable that the so-called lime-loving species are plants which prefer, not merely the lime, but the high mineral-content in general. However, in the absence of experimental work bearing specially on this point, no positive statements can be made; though Graebner (1909) bases a classification of the vegetation of Germany on an assumption, by no means an unreasonable one, which is nearly the same as this one.

TABLE I. CONSTITUENTS

	Aire Head Spring		Smelt Hill Water Sink	
	Grains per gallon	Millegrams[1] per litre	Grains per gallon	Millegrams[1] per litre
SiO_2	0·224	5·2	0·139	5·1
SO_3	1·050	38·6	2·049	75·5
HNO_3	0·100	3·4	—	—
HNO_2	—	—	—	—
P_2O_5	Minute traces	—	Minute traces	
Cl	0·700	25·8	0·700	25·8
Fe_2O_3	0·030	1·1	0·009	3·3
CaO	6·258	230·5	6·839	251·8
MgO	0·403	14·8	0·841	31·0
Na	0·456	16·8	0·456	16·8
NH_3	0·007	2·6	0·003	1·1
Albuminoid NH_3	0·006	2·4	0·018	6·6

TABLE II. DISSOLVED SALINE CONSTITUENTS

	Aire Head Spring		Smelt Hill Water Sink	
	Grains per gallon	Millegrams[1] per litre	Grains per gallon	Millegrams[1] per litre
SiO_2	0·224	8·2	0·139	5·1
$Ca(NO_3)_2$	0·130	4·8	—	—
$CaCO_3$	9·778	360·1	9·646	355·2
$CaSO_4$	1·785	65·6	3·483	128·3
$MgCO_3$	0·843	30·9	1·759	64·7
NaCl	1·155	42·5	1·155	42·5
$FeCO_3$	0·043	15·85	0·013	4·4
	13·958	513·9	16·195	550·6
Dissolved matter by evaporation	14·280	525·8	16·940	576·0

[1] 15·43 grains = 1 gram; 1·76 gallons = 1 litre.

SWAMPS (OR MARSHES) ON THE LIMESTONE

Swamps are much less numerous in the limestone tract than on the less porous sandstones and shales. Most of the swamps in the limestone area occur at the bottoms of the dales, although a few occur on the grassy slopes, wherever springs issue through the soil. A very large number of species which characterize the swamps of the sandstone and shale are absent from those of the limestone, whilst some species are restricted to the limestone swamps. The following species have been found in the limestone swamps of north Derbyshire; and it will be seen that the total floristic composition of the limestone swamps is very different from that of the sandstone swamps, just as the soluble mineral content of each is very different. These differences are sufficiently important to justify the placing of the two kinds of swamps in separate plant formations.

Thalictrum flavum	r	V. dioica	la
Trollius europaeus	la	Eupatorium cannabinum	la
Caltha palustris	a	Petasites ovatus	la
Chrysosplenium oppositifolium	la	Cnicus heterophyllus	a
C. alternifolium	la	C. palustris	a
Parnassia palustris	la	Festuca elatior	r
Spiraea Ulmaria	a	F. arundinacea	vr
Potentilla erecta	o	Scirpus compressus (Blymus compressus)	r
Geum rivale	la	Carex disticha	l
G. rivale × urbanum (=G. intermedium)	la	C. acuta	r
Epilobium hirsutum	a	C. flacca	o
Polemonium caeruleum	la	C. pendula	l
Myosotis palustris	la	C. strigosa	r
Mentha spp.	a	C. sylvatica	l
"M. rotundifolia"	r	Juncus glaucus	la
Pedicularis palustris	r	J. compressus	vr
Valeriana officinalis (=V. Mikanii)	la	Helleborine palustris	vr
		Orchis maculata	o to a

In similar limestone swamps of the mid-Pennines, the bird's eye primrose (*Primula farinosa*) is abundant, and there reaches its southern British limit.

Admitting that the two kinds of swamps should be placed in separate plant formations, the question arises as to which

Copyright *W. B. Crump*

Figure 21.

A larger Limestone Dale.

Calcareous grassland and screes on the hill slope on the left. Ash (*Fraxinus excelsior*) and Wych Elm (*Ulmus glabra = U. montana*) by the stream side. In the foreground the River Wye.

plant formations the two swamps respectively must be referred to. They can scarcely be placed in the aquatic formation, as they do not occur in water nor do many of the species exhibit marked hydrophytic characters. The plants of the swamps under discussion are not the "helophytes" of Warming (1909: 185), the latter being members of reed swamps, which are here included in the aquatic plant formation (see page 154). The plants of the siliceous and calcareous swamps here alluded to are obviously more terrestrial in character than the members of reed swamps, and should probably be placed in the two main formations of the district which, like the swamps in question, occur on the siliceous and calcareous soils respectively. On this basis, the swamps of the sandstones and shales would be placed in the same formation as the oak woods, scrub, and siliceous grassland; whilst the swamps of the limestone area would be placed in the same formation as the ash woods, scrub, and calcareous grassland. This arrangement conforms with the general edaphic conditions, the general floristic composition, and the topographical position of the two kinds of swamps in question (cf. the summary on p. 215). These swamps and their related vegetation are obviously allied to the "*Formation der Quellfluren*" of the Swiss plant geographers (cf. Rübel, 1911: 193).

Ruderal Marsh Species

In addition to the species which are definitely members of the swamps in the vicinity of springs, there are a number of other marsh plants of the district which cannot be referred to either of the preceding plant associations, and which in their distribution remind one of the ruderal plants inhabiting dry "waste places" and roadsides. Many of the marsh ruderal plants are found in wet places on stream-banks liable to be flooded in times of heavy rains; but they also occur in any kind of wet places to which their reproductive parts may happen to get carried. They are usually found in the more lowland localities, and they are more at home in the marsh associations of the alluvial plains. The names of a number of such plants are given on the next page; but, in addition to the following, several plants of the swamps also exhibit ruderal or viatical tendencies:—

Rumex conglomeratus	l	Mentha spp.	la
Polygonum Hydropiper	o	*Mimulus Langsdorfii	l
Ranunculus repens	a	Veronica Beccabunga	o to a
Nasturtium palustre	r	Gnaphalium uliginosum	o
N. officinale	la	Petasites ovatus	a
Barbarea vulgaris	o	Pulicaria dysenterica	r
Potentilla Anserina	a	Alopecurus geniculatus	o
Apium nodiflorum	r	Carex hirta	l
Sium erectum	r	Juncus bufonius	la
Scutellaria galericulata	o		

Reed Swamps

There being no extensive, shallow, sheets of water in the district, it follows that reed swamps are not common and are not typically developed. The most characteristic member (*Phragmites communis*) of reed-swamps in general has only, in the course of the present botanical survey, been observed in lower Monsal Dale, where, in the back waters of the river Wye, it forms one or two small associations or societies. However, most of the streams of this hilly district have a fairly strong current; and this factor of itself is sufficient to account for the paucity of reed swamps of *Phragmites communis* by the stream sides. It appears to be generally the case that this plant only forms typical reed swamps in stagnant or almost stagnant waters; and, where the water is well aërated, any reed swamps that occur are dominated by other plants, such as *Glyceria aquatica* and *Phalaris arundinacea*. In this district the two latter plants are frequently dominant in plant societies or small associations up to about 600 feet (183 m.). The latter plant, however, occasionally occurs up to about 900 feet (274 m.).

The following plants occur in the small and rather non-typical reed swamps of the southern Pennines: it will be seen that many of the species are in this district only of local occurrence:—

	In calcareous waters	In non-calcareous waters
Equisetum maximum	vr	vr
E. palustre	o	o
E. limosum	o	la
Polygonum amphibium	r	r
*Nasturtium amphibium	—	r
N. palustre	o	r
"Lythrum Salicaria"	vr	?
Epilobium hirsutum	a	la
Angelica sylvestris	o	o
*Lysimachia vulgaris	—	vr
Stachys palustris	o	o
Veronica Anagallis	o	—
Eupatorium Cannabinum	o	r
*Bidens tripartita	—	r
*Typha latifolia	—	vr
Sparganium ramosum	r to o	r to o
*S. simplex	vr	vr
Triglochin palustre	vr	vr
Alisma Plantago	r	r
*Sagittaria sagittifolia	vr	vr
Phalaris arundinacea	la	la
Agrostis alba	r to o	r to o
Phragmites communis	vr	vr
Glyceria aquatica	la	la
Eleocharis palustris	—	vr
Carex remota	vr	r
C. inflata (=C. ampullacea)	r	r
C. acutiformis (=C. paludosa)	r	r
C. vesicaria	vr	r
*Acorus Calamus	—	vr
Juncus effusus	l	la
J. inflexus	la	l
(=J. glaucus)	—	vr
Iris Pseudacorus	—	—

THE VEGETATION OF QUICKLY FLOWING STREAMS

("The hydrophyte formation round springs and streamlets."
Ostenfeld, 1908 : 943.)

The streams of the Pennine slopes are characterized by their quick current and their shallowness. A result of their quick current is that the water is well aërated; and this factor is related to the absence of a large number of lowland aquatic flowering plants. The shallowness of the streams assists towards the same result. The streams on the sandstones and shales are deficient in humous acids, except in the case of the streams which drain the peaty deposits of the plateaux. The streams of the limestones are not merely deficient in humous acids, but are actually alkaline in reaction.

On the flatter moorland plateaux in streams draining from the peat moors, a pond weed (*Potamogeton polygonifolius*) is indigenous. This plant appears to be confined to acidic waters. A crowfoot (*Ranunculus Lenormandi*) is locally abundant in similar situations, as well as in small streams and swamps on the hill-slopes of the sandstones and shales. *R. hederaceus* is very much rarer. *Glyceria fluitans* and *Callitriche stagnalis* are found in the streams, in reservoirs and mill dams, and in swamps up to 1500 feet (457 m.). *Montia fontana, M. minor*, and *Stellaria uliginosa* are also locally very abundant in the streams on the sandstones and shales. The Batrachian crowfoots are almost confined to the limestone streams, where, however, they are locally very abundant, especially so below 500 feet (152 m.). Species of Chara are also limited to the calcareous streams, whilst *Nitella opaca* occurs rarely in both the calcareous and the non-calcareous waters of the district.

Ostenfeld (*loc. cit.*) includes "the vegetation on cliffs over which water constantly trickles" with the vegetation of round springs and streamlets, and this appears quite a reasonable plan; and Ostenfeld (*loc. cit.*) is also quite reasonable in refusing to follow Jonsson in placing this type of vegetation with the moss-bogs described by Warming, which are characterized by stagnant and not by well aërated water.

On the whole, however, flowering plants are not generally abundant in quickly flowing streams; but this deficiency is more than compensated by the great abundance and variety of liverworts and mosses. Bog-mosses (*Sphagnum* spp.) never occur in the streams of the limestone hill-slopes, and are not common in those of the siliceous hill-slopes except when these are rich in humous acids. Algae are locally abundant and equally characteristic, but have not been studied ecologically.

The following liverworts and mosses have been observed in the streams of the southern Pennines:

	In calcareous waters	In non-calcareous waters
Hepaticae		
Jubula Hutchinseae	—	vr
Scapania undulata	—	a
S. purpurascens	—	l
"S. intermedia"	—	r
Chiloscyphus polyanthus	o	o
Jungermannia incisa	—	o
J. cordifolia	—	r
Nardia hyalina	—	r
N. obovata	—	r to o
N. compressa	—	o
"Pellia Neesiana"	—	l
Aneura multifida	r	r
A. pinguis	r	r
"A. ambrosioides"	—	r
Musci		
Sphagnum crassidulum	—	la
Dicranella Schreberi	r	r
D. squarrosa	—	o
Fissidens crassipes	r	—
Rhacomitrium aciculare	—	a
Cinclidotus fontinaloides	la	—
Orthotrichum rivulare	r	—
Aucolocomnium palustre	—	la
Philonotis fontana	—	la
"P. adpressa"	—	r
P. calcarea	r	r
Webera albicans	o	o
"Bryum filiforme"	—	r
Fontinalis squamosa	vr	la
F. antipyretica	r to o	la
"Porotrichum angustifolium"	vr	—
"Leskea polycarpa"	o	r
Brachythecium rivulare	r	la
B. plumosum	r	o
Hyocomium flagellare	—	o to a
Eurynchium rusciforme	a	a
Amblystegium irriguum	r	r
A. fluviatile	r	—
Hypnum riparium	r	o
H. uncinatum	r	o
H. fluitans (agg.)	—	la
H. commutatum	a	a
H. falcatum	o	o
H. palustre	o	o
H. ochraceum	o	a
H. stramineum	—	o

ALIEN AQUATIC PLANTS

A large proportion of the aquatic species occur in the canals and mill-dams, and were not recorded by the older botanists who flourished before these habitats were constructed. Hence it is unlikely that such plants existed as members of the primitive flora of a district which originally was almost if not quite destitute of natural sheets of still waters. However, on the construction of the canals and mill-dams about a century ago, aquatic plants speedily invaded the new habitats, thus proving that their previous absence from the district was only due to the absence of natural ponds and lakes, and not to any climatic reasons.

Most of the aquatic plants in question were not introduced here intentionally, but spread spontaneously; and these therefore belong to a different category from such plants as the white water-lily (*Nymphaea alba*) which has here and there been intentionally introduced into the ornamental waters of some of the parklands, as at Chatsworth.

Many of the canals and mill-dams which harbour these aquatic aliens are grossly contaminated with mill refuse; but a number of lowland aquatic species appear to be peculiarly unsusceptible to such influences, perhaps because they have become adapted to life in stagnant and often naturally foul and badly aërated waters. On the other hand, those aquatic plants which characterize swift-flowing streams, whose waters are naturally well aërated and pure, are speedily extirpated when the waters are rendered foul.

The invasion of the artificial stagnant waters by aquatic plants of the Pennines is still proceeding; and quite recently, an American pond-weed (**Potamogeton pennsylvanicus*) has become established in a canal near Halifax, a little to the north of the present district. Mr A. Bennett (1908 *a*) states that this is the first recorded instance of an alien Potamogeton becoming established; but all the Potamogetons of the canals which cross the Pennines are, in a sense, aliens in these upland waters.

In a canal near Manchester, a few miles to the west of the present district, such remarkable alien aquatic species occur as

**Chara Braunii*, **Naias graminea*, and **Vallisneria spiralis*. These, as has been recently shown, are confined to certain parts of the canal where the temperature is more or less permanently raised by the inflow of heated water from adjoining cotton mills (Weiss and Murray, 1909), and have no doubt been introduced with imported cotton. **Potamogeton pennsylvanicus*, which, when first discovered in the canal at Halifax, was thought to be restricted to the water which is heated by the discharge from cotton mills (Bennett, 1908 *a* and *b*), has been found to have a rather more extended range. This plant also has in all probability been introduced with cotton from North America.

It is evident, therefore, that many water plants have excellent means at their disposal for successful migration. Not only is this seen from the above-mentioned facts of distribution, but an experiment (Wheldon and Wilson, 1907 : 339) confirms this well-established opinion. A pond was made, near Garstang in North Lancashire, in a grassy field to see which water plants would appear. The pond was carefully railed off to prevent access of cattle. After eighteen months, the following plants had appeared:—*Nitella opaca*, *Callitriche* sp., *Alisma Plantago*, *Glyceria fluitans*, *Juncus conglomeratus*, and *J. articulatus*.

It may, perhaps, appear surprising that practically no alien aquatic plants should have become established on the Pennines in the numerous reservoirs (see figure 36) which have been constructed at the clough heads during recent years; but the reservoirs are artificially kept clear of water "weeds." However, the only recorded station of the water purslane (*Peplis Portula*) in the Peak District is on the south side of the reservoir at Chapel-en-le-Frith (Linton, 1903: 151).

The following is a list of the aquatic flowering plants of the ponds, mill-dams, canals, etc. of the southern Pennines. Most of the species occur only at altitudes below 600 feet (183 m.). Those which occur at the higher altitudes are indicated by the letter "U" (= upland) being placed in brackets after the specific name; and the plants that are not strictly indigenous are preceded by an asterisk (*):—

	In calcareous waters	In non-calcareous waters
Charophyta		
Chara fragilis	vr	—
C. hispida	vr	—
C. vulgaris	vr	vr
Nitella opaca	l	l
Angiospermae		
Polygonum amphibium	r	*vr
*Nuphar lutea	l	—
*Nymphaea alba	l	—
*Ceratophyllum demersum	—	vr
Ranunculus circinatus	vr	—
R. fluitans	la	—
R. pseudofluitans	la	—
R. trichophyllus	vr	—
R. Drouettii	vr	—
R. peltatus (agg.)	r	—
R. Baudotii	vr	—
"Callitriche verna" (U)	vr	*vr
C. stagnalis (U)	a	a
C. hamulata	vr	vr
C. obtusangula	vr	vr
Myriophyllum spicatum	r	*vr
*M. alterniflorum	—	vr
"Hippuris vulgaris"	vr	vr
*Apium inundatum	—	—
*Littorella juncea	—	—
Potamogeton natans	o	o
*P. perfoliatus	—	vr
P. crispus	vr	*vr
*P. obtusifolius	—	vr
*P. zosterifolius	—	—
*P. pennsylvanicus	—	vr
P. pusillus	l	l
*P. Friesii	—	*vr
*P. pectinatus	—	vr
Zannichellia palustris	vr	*vr
*Sagittaria sagittifolia	—	vr
*Elodea canadensis (U)	la	la
Lemna minor (U)	la	la
*L. trisulca	—	vr
*Heliocharis acicularis	—	vr
"Catabrosa aquatica"	vr	—
Glyceria fluitans (U)	a	a
G. plicata (agg.)	la	vr

The Relation of Mineral Salts to the Flora and Vegetation

It will be seen that the differences between the flora and vegetation of the calcareous and the non-calcareous terrestrial soils are paralleled by differences between the aquatic and sub-aquatic flora and vegetation. These differences cannot be explained by the differences in the physical nature of the habitat, for, in the case of the two aquatic habitats, the physical characters are identical. The facts therefore support the view that the presence of lime-loving plants in calcareous waters and soils is somehow related to the chemical composition of the waters (cf. Schimper, 1903: 102). Wheldon and Wilson (1907: 90) also point out that in addition to water-content, "another agent also has its influence. On examining the plant associations of habitats having about the same degree of humidity, we frequently find the species in them are totally dissimilar." These authors then go on to give the groups of species found respectively in boggy ground at the calcareous shore of a moorland tarn, with *Primula farinosa*, *Cladium Mariscus*, *Schoenus nigricans*, *Juncus obtusiflorus*, and *Hypnum falcatum*; of the swampy margin of a pool on the glacial drift, with *Ranunculus Flammula*, *Bidens tripartita*, *Alisma Plantago*, *Sparganium neglectum*, and *Hypnum aduncum*; of a peat bog, with *Drosera* spp., *Andromeda Polifolia*, *Carex limosa*, *Myrica Gale*, and *Sphagnum* spp.; and of estuarine marshes, with *Spergularia media*, *Limonium* (= *Statice*), *Aster Trifolium*, and *Glyceria maritima*. As "in all these stations the water supply is fairly constant and really in excess of plant requirements, and the conditions of altitude and exposure practically identical," the authors "are led, therefore, to the conclusion that some component of the soil must be the factor which determines the presence or absence of these various species"; and it seems to me also that a study of the distribution of plants in any district where the chemical composition of the soil shows great variations leads inevitably to the conclusion of Nageli (1865: 367), of Schimper (1903: 101), and many others that the chemical constituents of the soil, in addition to any or all of the physical conditions, have directly or indirectly an enormous influence on the differential distribution of the flora and vegetation of that district. Precisely what this influence is, however, is a matter on which the present state of knowledge in physiological botany allows no definite statement to be made.

Clements (1907: 18) opposes the "chemical theory" (cf. Schimper, *loc. cit.*); and even goes so far as to assert that "it now appears entirely incorrect to ascribe the presence or absence of certain species on limestone soils to the chemical nature of the latter." Without doubt, the "chemical theory" requires close investigation by modern plant physiologists; and a re-statement of the whole position is urgently needed. However, the study of the vegetation of a district like the present, where highly calcareous soils occur in close proximity with soils extremely poor in lime, where the climatic factors obtaining over the two types of soil are identical, where both habitats exhibit every transition from wet to dry, and yet where the flora and vegetation of the two types of habitat show very marked differences, should convince any observer that the "chemical theory" is far from being effete.

Hall and Russell (1911: 54) have recently stated that "soils devoid of carbonate of lime are never fertile, because without it the plant food of the soil cannot readily be brought into a condition available for the plant, and many of the most important bacterial actions in the soil are dependent on the presence of a base like carbonate of lime."

Kraus (1911) has recently published some valuable statistical data bearing on the question; and not until more work has been performed on these lines will it be possible to reach the solution of this intricate and much-debated problem.

CHAPTER VII

MOORLAND ASSOCIATIONS

General distribution of moorland. Classification of moorland associations. The fen formation. "Hochmoor" and "Flachmoor." Moors and Fens. Bryophytes of the moors. Factors related to the distribution of the moorland associations. Heather moors. Transitional association of heather moor and siliceous grassland. Bilberry moors. Transitional association of heather moor and bilberry moor. Cotton-grass moors. Transitional association of cotton-grass moor and siliceous grassland. Transitional association of heather moor and cotton-grass moor. Retrogressive moors. The Peak of Derbyshire. Bare peat. Sub-Alpine grassland. Zonation of the moorland and grassland associations. List of species of the moor formation.

GENERAL DISTRIBUTION OF MOORLAND

THE moorland associations occur on peat whose waters are poor in soluble mineral salts and are also acid in reaction. In this district, the peat is almost wholly confined to the plateaux and to the upper slopes of the sandstones and shales. Peat also occurs in places which the existing geological maps indicate as Carboniferous Limestone; but most of such localities on examination prove to have a surface layer of chert (figure 1). Certain volcanic and metamorphic rocks of the limestone area also bear moorland plants over limited tracts; and, as already stated (see pages 122—6), heather (*Calluna vulgaris*) and some of its associates are found on leached limestone soils. There are no lowland peat moors in this district, although they are not rare on the plains both on the east and west of the Pennines.

The peaty uplands consist of gently sloping plateaux. These are usually dip slopes, or less often the slopes of the valleys of small streams. The peat of the lower plateaux is shallow and usually dominated by heather (figures 33 and 34 *b*): the peat of the higher plateaux is much deeper and usually dominated by the cotton-grass (figure 33 *c*). This typical regularity of the moorland plateaux is, however, frequently broken by alternating outcrops of sandstone and shale. The outcrops of sandstone are usually characterized by a preponderance of bilberry (figure *d*), and those of the shale by swamps in which the larger rushes (*Juncus effusus* and *effusus* forma *compactus*) are generally conspicuous. The vegetation of these Juncus swamps has already been described (see pages 146—150). Sometimes, even in the general moorland area, a steep, shaly hill-slope is characterized by siliceous pasture.

The moorland vegetation ascends to the highest summits of the district, three of which attain an altitude slightly exceeding 2000 feet (610 m.). No Arctic or Alpine species occur anywhere on the southern Pennines, as in the case of the higher Pennine summit of Cross Fell further to the north (Lewis, 1904 *a*: 328; 1904 *b*: 279). The vegetation of the highest plateaux of the Peak District is marked by the occurrence of retrogressive moors (figures 28 to 31) with much bilberry (*Vaccinium Myrtillus*), which very often alternates with patches of crowberry (*Empetrum nigrum*), less frequently of cloudberry (*Rubus Chamaemorus*), and sometimes of bare peat.

The altitude to which the moorland vegetation descends is determined partly by physiographical and partly by artificial causes. Where the moorland ceases abruptly and is separated merely by a stone wall from the permanent pasture of the upland cultivation (see figure 24), the lower limit is simply the place where reclamation has stopped; and this limit usually varies from about 1000 feet (305 m.) to about 1600 feet (488 m.); but where the moorland adjoins siliceous pasture, the limit is a perfectly natural one. The latter limit varies from about 750 feet (229 m.) to about 1500 feet (457 m.). Where the moorland descends to the lower of these altitudes, a zone of heather moor invariably occurs; and where the moorland ceases at the higher of these altitudes, this zone is absent.

In ascending the lateral slope of an upland valley, the change from the grassy slope to the peaty plateau is rather abrupt, and the transition region narrow. This is well seen in an

 W. B. Crump

Figure 22.

Heather Moor on peat over Millstone Grit rock.
Heather (*Calluna vulgaris*) in flower in September.

ascent of the grassy hill sides north-west of Edale railway station, where the transition occurs at an altitude of about 1500 feet (457 m.). The transitional area is usually characterized by much bilberry and crowberry. On the other hand, in following the course of an upland stream to its source, the moorland plateau is reached very gradually; and, at the head of the valley, where a number of very different plant associations converge, a confused mixture occurs of siliceous pasture, scrub, and moorland—a mixture which defies accurate cartographical representation of the vegetation except on maps of a large scale. In descending a moorland plateau along the watershed between two lateral valleys, the moorland vegetation usually comes down to about 1000 feet (305 m.), and, in a few cases, as at Tintwistle Moor, near Glossop, to about 750 feet (229 m.).

The rocky, exposed summits of the higher hills (figure 25) are characterized by the dominance of the bilberry (*Vaccinium Myrtillus*). Such bilberry moors are not of great extent; but they are interesting as linking the vegetation of the Pennines with that of central Scotland, where bilberry moors at high altitudes are widespread (R. Smith, 1900 *b*: 461—2).

Sometimes the various moorland associations are sharply marked off from each other. Such sharply defined boundaries nearly always correspond with well-marked physiographical features. For example, a cotton-grass moor occupying a high plateau sometimes ceases quite sharply at an escarpment, on the plateau below which a heather moor may occur. The rocky escarpments, like the exposed rocky summits, are characterised by much bilberry; but the flora of the bilberry edges is richer than that of the bilberry ridges. Sometimes, however, the various moorland associations pass into each other very gradually, as when a heather moor adjoins a cotton-grass moor and there is not intervening escarpment. In such cases the transitional region is broad, and is marked by the co-dominance of the heather and the cotton-grass. On the accompanying vegetation maps, such transitional areas are indicated by stippling the colour used for heather moors on the colour used for cotton-grass moors.

Classification of Moorland Associations

The moorland plant associations of the district may be arranged and designated as follows:—

1. Heather moor. Association of *Calluna vulgaris* (Callunetum vulgaris).

2. Cotton-grass moor. Association of *Eriophorum vaginatum* (Eriophoretum vaginati).

3. Bilberry moor. Association of *Vaccinium Myrtillus* (Vacciniëtum myrtilli).

4. Retrogressive moors whose chief constituents are the bilberry, the crowberry (*Empetrum nigrum*), and the cloudberry (*Rubus Chamaemorus*). Locally, extensive patches of bare peat occur.

5. Transitional moors of heather and cotton-grass.

6. Transitional moors of heather and bilberry.

7. Transitional areas of heather and siliceous pasture.

8. Transitional areas of cotton-grass and siliceous pasture.

In other parts of the British Isles, the following additional moorland plant associations have been recognised.

9. Sphagnum moor. Sphagnum moors are mapped by Lewis (1904*a*: 325). They are also described for north Lancashire by Wheldon and Wilson (1907: 102) who state that this "upland Sphagnum association" has for its dominant species *Sphagnum recurvum*, and that this is occasionally accompanied by other bog mosses, such as *S. nitens*, *S. papillosum*, and *S. rubellum*, and still more frequently by *Polytrichum commune*. Frequent members of this association, though usually marginal or very subordinate, are *Viola palustris*, *Vaccinium Oxycoccus*, *Juncus effusus*, *Eriophorum angustifolium*, *Carex echinata*, *C. canescens*, and locally *Rhyncospora alba*. One of the Sphagnum moors mapped by Lewis (*op. cit.*) has more recently been visited by the members of the Yorkshire Naturalists' Union (see *The Naturalist*, 1910: 265 and 313), who reported, in addition to many other mosses and Hepatics, the following species of Sphagnum:—

S. rubellum, S. acutifolium, S. subnitens, S. cuspidatum, S. recurvum, S. inundatum, S. tenellum.

10. Cotton-grass moor. Association of *Eriophorum angustifolium* (Eriophoretum angustifolii).

11. Scirpus moor. Association of *Scirpus caespitosus* (= *Trichophorum caespitosum*) (Scirpetum caespitosi).

These two associations are described by Pethybridge and Praeger (1905: 166—7), and occur on the Wicklow mountains south of Dublin. In certain Scottish localities also, namely, in the north-western Highlands, in the Hebrides, and in the Shetlands, "*Scirpus caespitosus* dominates the most characteristic associations. In N.W. Sutherland, the flat, basaltic plateau of northern Skye, and the valley floors and gently sloping hillsides of Shetland, this type remains dominant over many hundred square miles.... In the western portion of N. Uist, the *Scirpus caespitosus* association extends over hills of low elevation and broad shallow valleys" ([F. J. Lewis, in] *The New Phytologist*, 1908: 257). Scirpus moors resemble the cotton-grass moors and heather moors of the Pennines in the comparative scarcity of bog-mosses (*Sphagnum* spp.).

12. Rhacomitrium moor. Association of *Rhacomitrium lanuginosum* (Rhacomitriëtum lanuginosi). Pethybridge and Praeger (1905: 167) describe a modification of the heather moor and the Scirpus moor brought about by the local dominance of a species of woolly moss (*Racomitrium lanuginosum*). The association of *Rhacomitrium lanuginosum* would appear to be fairly widespread in western Scotland; and "Rhacomitrium bogs" in Caithness are described by Crampton (1911: 55).

All the above associations and societies occur on peat which is marked by acidic waters, and by a low soluble mineral-content, especially by a low lime-content. The plant formation of such a habitat has been termed an Oxodion (Moss, 1910 *b*: 43); and its described chief associations in the British Isles may be designated as follows (cf. Moss, *loc. cit.*):—

Oxodion Calluneti-vulgaris
Oxodion Eriophoreti-vaginati
Oxodion Vacciniëti-myrtilli
Oxodion Sphagneti-recurvi
Oxodion Eriophoreti-angustifolii
Oxodion Scirpeti-caespitosi
Oxodion Rhacomitriëti-lanuginosi

The Fen Formation

In the lowlands of eastern England, peat occurs which is characterized by alkaline waters, and by a high, soluble mineral-content, especially by a high lime-content. Such peat bears a totally different set of plant associations and societies and quite a different flora; and its vegetation must therefore be placed in a different formation. The peaty tracts of eastern England which are characterized by alkaline waters are usually spoken of by the local inhabitants as "black fens" or simply "fens"; and the difference between the two types of peaty vegetation appears to be of a very fundamental nature. Accordingly, the vegetation of acidic peat may be said to belong to the moor formation and that of alkaline peat to the fen formation.

"Hochmoor" and "Flachmoor"

Continental plant geographers distinguish two main types of peat vegetation under the names of "Hochmoor" and "Flachmoor." In Warming's *Oecology of Plants* (1909: 204), these terms are represented by "high moor" and "low moor" respectively. These distinctions do not entirely correspond to the distinctions of moor and fen adopted by most British plant geographers. Warming (*loc. cit.*) gives nine distinctions between "high moors" and "low moors."

1. "Low moor arises on a surface that is covered with water....High moor arises on moist soil or even above water." On this basis, the moors of the Peak District are "high moors."

2. "Low moor has a flat surface (either horizontal or inclined). High moor has a convex surface." The cotton-grass moors of this district have, on the whole, a flat surface; and, therefore, if judged from this point of view alone, they would have to be placed among "low moors."

3. "Low moor is produced particularly by grass-like plants, including Cyperaceae....High moor owes its origin to bog-mosses, Sphagnum and others, and includes many Ericaceae." Applying this test, the heather moors would be placed among the "high moors," the cotton-grass moors among the "low moors"; but this test for low moors is unsatisfactory, as Warming (p. 202) also includes "grass-like plants," such as *Rhyncospora alba*, *Carex* spp., *Eriophorum* spp. (especially *E. vaginatum*), and *Agrostis canina*, among the constituent plants of "high moor."

Copyright *W. B. Crump*

Figure 23.

Heather Moor.

Thin peat over Chert on the Limestone Plateau. Associated with the Heather (*Calluna vulgaris*) are Hawthorn (*Crataegus Oxyacantha* = *C. monogyna*) and dwarf Furze (*Ulex Galii*). In the background is High Rake and a slope of upland calcareous grassland.

4. "Low moor water is calcareous. High moor water contains little or no lime." Judged by this test, which seems a good one, all the moors of the Peak District are "high moors." However, some moors placed by some continental phyto-geographers among Flachmoors are characterized by species which grow on peat whose water contains little or no lime.

5. "Low moor forms black, amorphous peat....High moor preserves its plants in a higher degree." From this standpoint, all the moors of this district are "high moors."

6. "Low moor peat is heavy and rich in mineral bodies (with ten to thirty per cent. of ash). High moor peat is light in weight and poor in mineral matter (with about five per cent. of ash)." The peat of this district yields much less than five per cent. of ash, if silica (SiO_2) be excluded; and the application of this test therefore may be regarded as placing the moors among "high moors." Peat from Wicken Fen, near Cambridge, on the other hand, yields more than ten per cent. of ash.

7. "Low moor peat is usually close in texture,...and conducts water badly....High moor peat...conducts water well." The upper layers of the peat of all the moors of this district conduct water well.

8. "Low moor peat is very rich in food-material...High moor peat...is very poor in nutriment." The peat of this district is invariably poor in food-material.

9. "On low moor, mycorhiza and carnivorous plants are rare....On high moor, mycorhiza and carnivorous plants are common." On the moors of this district, the only carnivorous plants to be met with are the sundew (*Drosera rotundifolia*) and the butterwort (*Pinguicula vulgaris*); and both are rare. Mycorhiza occur in the roots of heather (*Calluna vulgaris*), but have not been proved to be present in many other moorland plants of the district. As carnivorous plants (e.g., *Utricularia minor*, *U. intermedia*, and *U. vulgaris*) occur in the waters of fens, this test is not very satisfactory.

Though some of these tests require re-stating, it seems fairly clear that the moors of the district belong to the class of Warming's "high moors"; and it is clear that some moors designated "Flachmoors" by continental plant geographers require re-investigation from the standpoint of the amount of available food-material contained in the peat.

Tansley (1911: 208) has recently discussed the conditions of British moors and fens; and in his *Types of British Vegetation*, accounts are given of the vegetation of the two plant formations. The "fenland formation" of Caithness, recently described by Crampton (1911: 74), is certainly not true fen, but a type of vegetation intermediate between fen and moor, and termed *Uebergangsmoor* (transitional moor) by Weber (1908: 95).

Moors and Fens

The only test of fen peat and moor peat, which is here regarded as really fundamental, is that depending on the amount of soluble mineral matter in the peat; and as this is very low in the case of all the local peats examined, there is no difficulty in referring all the plant associations developed on peat in this district to the moor formation as opposed to the fen formation.

The following characteristics distinguish the two formations:—

1. Fen peat is rich, moor peat is poor in soluble mineral matter.
2. Fen waters are alkaline, moor waters acid in reaction.
3. Fen peat often, moor peat rarely, contains the remains of molluscan shells.
4. The following plants are locally subdominant or very abundant on the peaty fens of eastern England:—

Cladium Mariscus
Schoenus nigricans
Phragmites communis
Calamagrostis lanceolata
Molinia caerulea
Juncus obtusiflorus

Of these species, only one—*Molinia caerulea*—occurs on the moors of the Peak District.

The following species are locally dominant or very abundant on British moors:—

Sphagnum spp.
Polytrichum commune
Rhacomitrium lanuginosum
Empetrum nigrum
Rubus Chamaemorus
Erica cinerea
E. Tetralix
Calluna vulgaris
Vaccinium Myrtillus
V. Vitis-idaea
Scirpus caespitosus
Eriophorum vaginatum
E. angustifolium
Carex Goodenowii
Molinia caerulea
Juncus squarrosus

Of these species, only *Molinia caerulea* and perhaps *Scirpus caespitosus* occur in the fen formation of Cambridgeshire.

5. The following species also occur (or formerly occurred) in the fens of East Anglia: those that are local or very rare are marked by an obelisk:—

Lathyrus palustris	C. paradoxa
† Viola stagnina	C. lasiocarpa
† V. montana	C. lepidocarpa
Peucedanum palustre	C. pseudo-Cyperus
† Selinum caruifolium	Calamagrostis Epigejos
† Senecio palustris	† Luzula pallescens
† S. paludosus	Orchis incarnata
† Sonchus palustris	Epipactis palustris
Potamogeton plantagineus	Habenaria conopsea
Carex disticha	† Liparis Loisellii

The following plants are characteristic of British moors, and are all common or fairly common plants:—

Lycopodium Selago	Rhyncospora alba
Blechnum spicant	Carex curta
Viola palustris	C. echinata
Drosera spp.	C. binervis
Galium saxatile	Agrostis canina
Arctostaphylos spp.	Nardus stricta
Vaccinium Oxycoccus	Deschampsia flexuosa
Andromeda Polifolia	Luzula multiflora
Myosotis repens	Narthecium ossifragum
Pedicularis sylvatica	Listera cordata
Potamogeton polygonifolius	Orchis ericetorum

An opinion which is held by many geologists and others to the effect that moor peat is generally composed of Sphagnum-moss and fen peat of Hypnum-moss is not supported by an examination of the plant remains preserved in the peat.

Certain types of vegetation, intermediate in various respects between moor and fen, require further study. They are usually characterized by the presence of some of the following species:—

Myrica Gale	Pedicularis palustris
Ranunculus Lingua	Schoenus nigricans
Potentilla Comarum	Triglochin palustre
Parnassia palustris	Eleocharis acicularis
Anagallis tenella	Scirpus pauciflorus
Menyanthes trifoliata	Eriophorum latifolium
Veronica scutellata	Orchis latifolia

BRYOPHYTES OF THE MOORS.

The cryptogamic flora of the various British plant associations has not yet been fully investigated. During the course of the present botanical survey, lists have been compiled of mosses and liverworts; but the Algae and the Fungi have not been fully investigated. I have frequently been indebted to Mr C. Crossland, of Halifax, for help in the identification of the liverworts and mosses; and I have also found the list of mosses in the floras by Linton (1903) and Crossland (1904) of very great service. The following mosses and Hepatics occur in the moor formation of the southern Pennines:—

Hepaticae

Blepharozia ciliaris
Lepidozia reptans
L. setacea
Kantia Trichomonis
Cepalozia lunulaefolia
C. bicuspidata
C. Lammersiana
C. divaricata
Scapania irrigua
S. nemorosa
Mylia anomala
M. Taylori
Jungermannia inflata
J. sphaerocarpa
J. crenulata
J. ventricosa
J. incisa
J. gracilis
J. lycopodioides
Nardia scalaris

Musci

Sphagnaceae

S. fimbriatum
S. rubellum
S. acutifolium
S. subnitens
S. molle (rare)
S. squarrosum
S. teres (rare)
S. compactum
S. rubsecundum
S. inundatum
S. Gravetii
S. rufescens
S. crassicladum
S. turfaceum
S. cuspidatum
S. recurvum
S. cymbifolium
S. papillosum

Polytrichaceae

Polytrichum urnigerum
P. nanum (rare)
P. piliferum
P. juniperinum
P. gracile
P. commune

Copyright *W. B. Crump*

Figure 24.

Heather Moor.

The moor abuts on upland permanent pasture.

Other mosses

Tetraphis pellucida	Leptodontium flexifolium
Dicranella crispa	Splachnum sphaericum (on dung)
D. cerviculata	Aulacomnium palustre
D. heteromalla	Webera nutans
Campylopus flexuosus	Bryum pallens
C. pyriformis	Mnium subglobosum
C. fragilis	Hypnum fluitans
Dicranum Bonjeani	H. exannulatum
D. scoparium	H. revolvens
Leucobryum glaucum	H. falcatum
Rhacomitrium spp. (rare)	H. stramineum

Factors related to the Distribution of the Moorland Associations

Factors which appear to be of importance in determining the various plant associations of the moors are (1) the relative amounts of sand and humus in the soil, (2) the amount of water in the soil, (3) the depth of the peat, (4) the altitude above sea-level, (5) the exposure to winds, (6) the anatomical structure of the moorland plants, (7) a change in the nature of the habitat, and perhaps (8) rainfall.

(1) Sand and humus. The peat of the heather moors yields a much greater quantity of silica (SiO_2) than the peat of the cotton-grass moors.

(2) Water. The peat of the heather moors possesses a much lower physical water-content than the peat of the cotton-grass moors; and this fact illustrates the general rule that the more humus a soil contains the more water it also contains.

It would appear that the insoluble particles of silica are of some importance in controlling the water-content, and thus of importance in distinguishing the various associations. Correlated with the water-content is the aëration of the peat; and the peat of the heather moor is much better aërated than that of the cotton-grass moor.

(3) Depth. The peat of the heather moor is shallower than the peat of the cotton-grass moor. That of the heather moor varies from a few inches to about four or five feet (122 or 152 cm.), and is commonly about a foot (30·5 cm.) in depth. That of the cotton-grass moor varies from about three feet

(91 cm.) to about fifteen feet (457 cm.): commonly it is about ten feet (30·5 cm.) deep; whilst locally in hollows these depths may be exceeded. The lower layers are almost constantly wet, and hence act as an impervious substratum to the upper layers which, however, are sometimes very dry in summer owing to evaporation.

(4) Altitude. The heather moor rarely exceeds 1500 feet (457 m.) in altitude: the cotton-grass moors ascend to 2000 feet (610 m.). Between 1500 feet and 1750 feet (533 m.), heather and cotton-grass frequently share dominance.

(5) Exposure. The exposed ridges and peaks, from 1500 feet upwards, are characterized by an association of bilberry (*Vaccinium Myrtillus*), whilst on the highest plateaux, retrogressive moors occur.

In general, it may be said that heather moors are found in the drier, more sandy, shallower, and less elevated regions, that cotton-grass moors dominate the wetter, purer, and deeper peat at higher elevations, that bilberry moors occur on the highest and most exposed ridges, and that the natural drainage resulting from the denudation of the peat of the cotton-grass moors on the highest plateaux and watersheds produces the associations characterized by *Vaccinium Myrtillus*, *Empetrum nigrum*, and *Rubus Chamaemorus*.

(6) Structure. The moorland plants possess certain morphological or structural peculiarities which enable them to thrive in their respective surroundings.

The vegetative organs of *Eriophorum vaginatum* and *E. angustifolium* and many of their associates are well provided with aëration canals which enable the underground parts to respire although they are embedded in peat which is, during the greater part of most years, supersaturated with water. Such plants are neither complete xerophytes nor complete hydrophytes, but possess both xerophilous and hydrophilous characters. This peculiarity of moorland plants has been pointed out by Warming (1896: 177). They are frequently termed "bog xerophytes" or "swamp xerophytes" (see also Yapp, 1909: 275—6).

The root-systems of *Calluna vulgaris*, *Vaccinium Myrtillus*, and *Empetrum nigrum* are superficial; and these plants have no aëration canals. These facts seem to be obviously related to the drier habitats of these species as compared with the habitat of the cotton-grasses (*Eriophorum* spp.) and their ecological allies. In the transitional moors of heather and cotton-grass, the shallow roots of *Calluna vulgaris, Vaccinium Myrtillus,* and *Empetrum nigrum* allow of their growth side by side with *Eriophorum vaginatum* whose functional roots are more deeply embedded in the peat; for, in summer and autumn, the upper layers of peat are frequently dry whilst the lower layers remain extremely wet.

(7) Changes in the habitat. It has already been stated that the peat on the highest moors is in a state of denudation and now dominated by *Vaccinium Myrtillus, Empetrum nigrum,* and *Rubus Chamaemorus.* An examination of the plant-remains composing the peat of such associations proves that the vegetation was previously dominated, and almost exclusively dominated, by Eriophorum. The degeneration of an Eriophorum moor results in the peat becoming drier; and this results in the dying out of the more hydrophilous species, such as *Eriophorum vaginatum* and *E. angustifolium* and the successful invasion of other plants, such as the bilberry, which are structurally better adapted to the drier conditions.

(8) Rainfall. It would appear that the local differences in the mean annual rainfall of the different parts of the moorland area have little or no relation to the local distribution of the different moorland associations in the Peak District. Generally speaking, the mean annual rainfall of the moorland area varies from about 40 inches (102 cm.) per annum to 55 inches (140 cm.) or rather more. It has been suggested (Smith and Rankin, 1903: 155) that the areas where the heather (*Calluna vulgaris*) is dominant have, on the whole, a lower mean annual rainfall than the areas over which the cotton-grass (*Eriophorum vaginatum*) is dominant; but, judging from the statistics supplied by Dr H. R. Mill (see page 25), the suggestion does not appear to be a fruitful one so far as this district is concerned. It is well known that the highest local rainfall of a district is not exactly at the summit of a hill, but some distance to the leeward of that hill. For example, in the present district, the highest indicated rainfall (see *British Rainfall*) is not on the Peak itself, but in Fairbrook Clough, which is a few miles to the leeward, that is, to the north-east of the Peak; and this tract of highest

local rainfall is characterized by a well-developed and typical heather moor. Again, whilst the vegetation maps of districts north of the Peak might, if taken by themselves, be held to indicate that the heather moors were characteristic of the eastern and not of the western Pennines, it will be seen that that indication is not borne out by the vegetation maps of the present district. The comparative dryness of the peat of the heather moors is to be explained, not by rainfall statistics, but by the larger proportion of silica mixed with the peat of the heather moors, and by the greater shallowness of the peat.

Heather Moors

Associations dominated by the common heather (*Calluna vulgaris*) are among the most typical plant associations of the British Isles. In a general way, such associations may be subdivided into heaths and heather moors, the former occurring on soils containing a higher proportion of sand and which are therefore drier, and the latter on soils containing a higher proportion of acidic humus and which are therefore wetter. In general, heaths are characteristic of the south and east of Britain, heather moors of the north and west. Hence, the occurrence of heath and of heather moor in this country would seem to be determined to some extent by climate. The associations of *Calluna vulgaris* (see figure 22), which so often occur as a fringe of the Pennine peat moors, are, on the whole, heather moors and not heaths, though some of those found at lower altitudes approximate in character to heaths.

Weber (1908: 91) suggests that the term moor should be used only when the peat is 20 cm. or more in thickness and when there is less than forty per cent. of ash [including silica] in the peat; but a too rigid use of these criteria lead to an artificial classification.

Beginning at Hayfield, what may be called the western system of heather moors extends northwards for about ten miles (16·1 km.), their continuity being broken by the narrow but deep clough formed by Shelf Brook, and by the larger Longdendale, formed by the river Etherow. The northern slopes of

Longdendale are, except at the extreme west, destitute of heather moors; but the main western Calluna mass runs along the southern slopes of the dale for six miles, as far as Woodhead. The most westerly outlier of the heather moors of the Pennines in this latitude occurs at Bakestone moor one mile to the west of the boundary of the district. On the Yorkshire slopes of the hills, beginning at Dunford Bridge, the eastern system of heather moors extends in a general south-easterly direction for about sixteen miles, and is continued eastwards of the present area on to the Sheffield map, no botanical survey of which has been completed. The central system of heather moors occupies a region in the upper portions of the valleys formed by the rivers Derwent and Westend. The eastern heather moors are about three miles broad on the average, and the western about one mile broad. This is a response by the vegetation to the well-known physiographical fact that the eastern slopes of the Pennines descend more gradually into the plain than the westerly slopes. The local altitudinal limit of the association, at about 1500 to 1550 feet (457 to 472 m.), is partly a response to the severer climatic conditions of the higher and more exposed summits, and partly to the wetter soil conditions which obtain on the deeper peat of the higher moors. The latter fact is doubtless related in part to the higher rainfall and more frequent mists which occur in these regions. It need scarcely be stated that the upper limit of the moorland association of *Calluna vulgaris* in no way corresponds with the upper limit of the species, which, as a matter of fact, ascends to over 2000 feet (610 m.) in this district (*e.g.*, on Bleaklow Hill); whilst in Scotland (Hooker, 1884) the species ascends to 3500 feet (1067 m.).

Many outliers or detached areas of heather moors occur, and are interesting as pointing to a former greater extension of the region of heather moors, a region which has been greatly restricted by reclamation and conversion into farmland. The sides of the roads and lanes in such reclaimed areas are frequently tenanted by moorland plants, such as *Calluna vulgaris*, *Vaccinium Myrtillus*, and *Deschampsia flexuosa*.

The western Pennines in the north of the Glossop district are remarkably destitute of heather moors: this is partly due to the fact that the slopes of the hills in that locality are very steep and shaly.

The vegetation of the roadsides, the footpaths, the banks of streams, and the fringe of the heather moors is, in general, of a grassy nature. In winter, it is possible to distinguish at a distance of some miles the sinuous course of old footpaths and bridlepaths by the contrast in colour which the bleached haulms of the mat-grass (*Nardus stricta*) make with the surrounding dark-coloured moorland vegetation.

Ostenfeld (1908: 887) states that the heather moors of the Faeröes are always met with on slopes with a southern exposure: this is not the case with regard to this association on the Pennines; and the inference is probably to be made that in the Faeröes, the association of heather moor is near its climatic limit of distribution.

On some of the more lowland of the heather moors, especially in sheltered depressions, the bracken (*Pteris aquilina*) is very abundant. It is possible that this plant is extending its range on the moors. Wheldon and Wilson (1907: 104) state that on the heather moors of North Lancashire, "where grouse are a prime consideration, the bracken is mowed periodically at considerable expense, and the heather is then enabled to overcome all rivals."

There is not a great deal of human interference with nature on the heather moors, although they are systematically fired by the keepers every few years. The length of time which elapses between the periods of firing varies locally, and determines the height to which the heather grows. On Eyam Moor the heather is fired about every four years, and therefore does not grow much more than ankle high. On the remoter moors in upper Derwent Dale, a period of eight to ten years elapses between the periods of firing; and the heather, on some of these moors, is frequently more than knee deep. For one or two years after the heather has been fired, the heather moor presents a desolate appearance; for the heather does not strongly reassert itself until at least two years have elapsed. The first plant to become conspicuous after firing is the bilberry (*Vaccinium Myrtillus*). The latter frequently occurs as a partially etiolated plant under the dominant heather, where its habit simulates that of *Listera cordata*. In such circumstances, however, the bilberry rarely flowers or fruits. The underground stems and buds of the bilberry are frequently unharmed by the firing, even when the

Copyright *W. B. Crump*

Figure 25.

Bilberry Moor.

Crest of a hill (1700 feet: m.) occupied by Bilberry (*Vaccinium Myrtillus*).

heather (which has no underground buds) has been completely killed. *Deschampsia flexuosa* and *Nardus stricta* are also frequently conspicuous during the first summer after the firing, doubtless owing to invasion by seed. As a rule, seedlings of heather establish themselves immediately and in abundance after firing; and, when this occurs, the complete and speedy rejuvenation of the heather moor is assured. The repopulating of the moor by heather is due to the germination of its minute seeds which are blown from adjoining heather-clad tracts. This fact is known to the keepers who therefore do not fire large, continuous areas in any given year. Seedlings of heather (*Calluna vulgaris*) may be found in abundance in places where the moor was burned during the previous year.

Typical dry heath, which is characteristically developed on sandy soils throughout the lowlands of England and especially so in the south and east, does not occur on the Pennines. Graebner (1901) has shown that this association does not in North Germany occur in localities where the rainfall is below 28 inches (71 cms.); and it would also appear, judging from its distribution in England, not to be developed where the mean annual rainfall is above 35 inches (89 cms.).

The following species occur in the less wet, that is, the typical parts of the heather moors of the southern Pennines:—

Dominant

Calluna vulgaris

Locally sub-dominant

Erica cinerea
Vaccinium Myrtillus

Locally abundant

Polytrichum spp.
Pteris aquilina
Ulex Gallii
Empetrum nigrum
Vaccinium Vitis-idaea
Galium saxatile
Deschampsia flexuosa
Juncus squarrosus

Occasional

Cladonia spp.
Lecanora sp.
Dicranum scoparium
Campylopus flexuosus
Webera nutans
Plagiothecium undulatum
Blechnum Spicant
Potentilla erecta
Calluna vulgaris *var.* Erikae
Molinia caerulea
Festuca ovina
Nardus stricta
Scirpus caespitosus
Pyrola media

Local, rare, or very rare

Lycopodium clavatum
Nephrodium dilatatum
Salix repens
Betula pubescens
Quercus sessiliflora
Rumex Acetosella
Crataegus Oxyacantha
Pyrus Aucuparia
Genista anglica
Lathyrus montanus
Polygala serpyllacea
Ilex Aquifolium
Calluna vulgaris *forma* incana
Trientalis europaea
Melampyrum pratense (agg.)
"Antennaria dioica"
Agrostis tenuis
Aira praecox
"Carex dioica"
G. Goodenowii
C. flacca
C. pilulifera
C. binervis
Luzula multiflora
"Listera cordata"

Locally, in damp hollows or wherever the soil is wet and badly aërated, the following additional species may occur:—

Sphagnum spp.
Polytrichum commune
Hypnum spp.
Lycopodium spp.
Ranunculus Flammula
  forma radicans
"Drosera anglica"
D. rotundifolia
"Potentilla palustris"
Viola palustris
Hydrocotyle vulgaris
Andromeda Polifolia
Erica Tetralix
Vaccinium Oxycoccus
Pedicularis sylvatica
"P. palustris"
Pinguicula vulgaris
Cirsium palustre
Agrostis canina
Eriophorum vaginatum
E. angustifolium
"Scirpus pauciflorus"
Carex echinata
C. curta
C. Goodenowii
  var. juncella
C. flacca
C. panicea
C. flava
Juncus acutiflorus
Narthecium ossifragum
Orchis ericetorum

Copyright *W. B. Crump*

Figure 26.

Cotton-grass Moor.

Cotton-grass (*Eriophorum vaginatum*) in fruit in June.

Transitional Association of Heather Moor and Siliceous Grassland

Some of the uncultivated areas are intermediate between heather moor and siliceous grassland. Most commonly in such places one finds patches of rough grass, such as the mat-grass (*Nardus stricta*), silver hair grass (*Deschampsia flexuosa*), and bent grass (*Agrostis tenuis*) alternating with patches of heathy dwarf shrubs, such as heather (*Calluna vulgaris*), bilberry (*Vaccinium Myrtillus*), and the fine-leaved heath (*Erica cinerea*). At other times, one finds a closer but dwarfed growth of the heathy dwarf shrubs but with a much larger proportion of the grasses than occurs on heather moor. It is rather rare to find the cowberry (*Vaccinium Vitis-idaea*) and the crowberry (*Empetrum nigrum*) in this transitional area; but the combination does occur. The soil is not peaty, although it contains more humus (of the acidic type) than is found on siliceous grassland. Vegetation of this nature is found not infrequently on the lower sandstone plateaux; and it is quite uncommon on the steep, shaly slopes. This is a subordinate association and probably of progressive nature; and it is here regarded as a stage in the succession from siliceous grassland to heather moor. On the vegetation maps, it is indicated by stippling the red colour used for heather on the colour used for siliceous grassland.

Plants from the neighbouring plant associations and even plants from the cultivated area invade the heather moor, especially along the footpaths and streamsides; and, in such places, the following species sometimes occur mixed with the heather:—

Urtica dioica (local)
Rumex Acetosa
Ranunculus bulbosus
Viola Riviniana
V. lutea (local)
Polygala serpyllacea
Heracleum spondylium (local)
Plantago lanceolata
Achillaea Millefolium
Bellis perennis
Hypochaeris radicata
Hieracium Pilosella
Taraxacum officinale
 var. maculiferum
Agrostis tenuis
 (=A. vulgaris)
Aira praecox
Triodea decumbens
Luzula campestris
L. vernalis
 (=L. pilosa)

Bilberry Moors

On the precipitous faces of the dry exposed sandstone rocks, very few plants occur. A few lichens, especially *Parmelia saxatilis*, occur; and such situations furnish the few remaining stations of species of the rock-mosses (*Andreaea* spp.). Sand and peat, however, find a lodgment on the rock-ledges; and here a few moorland plants, especially the bilberry (*Vaccinium Myrtillus*), the cowberry (*V. Vitis-idaea*), and the hair-grass (*Deschampsia flexuosa*), and even a few trees, such as dwarfed examples of birch (*Betula pubescens*, and *B. pubescens* var. *parvifolia*), oak (*Quercus sessiliflora*), and mountain-ash (*Pyrus Aucuparia*), find a home. On the screes and boulder-strewn slopes at the foot of the cliffs, the bilberry and the cowberry are often very abundant, as well as the crowberry (*Empetrum nigrum*) and the bearberry (*Arctostaphylos Uva-ursi*).

Such Vaccinium associations are exceedingly characteristic and very largely developed on the rocky slopes surrounding the Peak (see figure 20).

The bilberry also becomes dominant on the high, bleak, and wind-swept ridges and peaks of the sandstone hills. Such Vaccinium ridges have been described by Smith and Moss (1903: 381), and by Smith and Rankin (1903). As regards the Peak District, a typical Vaccinium crest is crossed by the public footpath going from Hayfield to the Snake inn. There is not very much difference in floristic composition between Vaccinium crests and Vaccinium edges, but on the former the cotton-grass and the cloudberry are often abundant, and the stations of the bearberry seem to be confined to the latter. If we regard the Vaccinium crest as an association ("Hauptypus"; Schröter, 1902), then the Vaccinium edge would be a sub-association ("Nebentypus"; Schröter, *op. cit.*).

The following is a list of the characteristic plants of a Vaccinium edge:—

Dominant

Vaccinium Myrtillus

Locally sub-dominant

Calluna vulgaris
Vaccinium Vitis-idaea

Copyright *W. B. Crump*

Figure 27.

Junction of Heather Moor and Cotton-grass Moor.

The dark patches are Heather (*Calluna vulgaris*), occupying the better drained parts of the moor. The other vegetation consists of tufts of Cotton-grass (*Eriophorum vaginatum*).

Locally abundant

Pteris aquilina	Erica cinerea
Empetrum nigrum	Galium saxatile
Arctostaphylos Uva-ursi	Deschampsia flexuosa

Occasional or rare

Betula pubescens	Pyrus Aucuparia
Quercus sessiliflora	Crataegus Oxyacantha
Rumex Acetosella	(=C. monogyna)

Very rare or extinct

"Andreaea alpina" "A. petrophila"
"A. crassinervia"

Transitional Association of Heather Moor and Bilberry Moor

The areas that are intermediate between heather moor and bilberry moor are shown on the map by dotting the red colour used for heaths over the purple colour used for bilberry. Such areas are usually rocky and peaty, like all the grounds characterized by stable bilberry moors; but they occur, as a rule, at rather lower altitudes than the latter.

Cotton-grass Moors

Cotton-grass moors occur on the gently sloping plateaux at elevations varying, as a rule, from about 1200 feet (363 m.) to 2000 feet (610 m.). These moors are locally termed "mosses"; and the place-name "moss," meaning a morass, is by far the most abundant place-name on the Pennine summits. Smith and Moss (1903: 380) and others have therefore used the name "moss moor," reminding one of the German "moosmoor," for this plant association. The place-name "moss," originally of physiographical significance, has provided the local plant-name for the chief constituent of the moor whose dominant plant (*Eriophorum vaginatum*) is well known to the inhabitants of the moor-edges as "moss-crops."

The peat of the cotton-grass moors is frequently ten to

fifteen feet (305 to 457 cm.) in depth, and rarely less than five feet (152 cm.). On rare occasions, as in local hollows and swamps, it may reach a depth of twenty feet (610 cms.) or more. The peat is usually saturated and frequently supersaturated with water, although the superficial layer occasionally becomes very dry in summer.

Over many parts of the higher moors, *Eriophorum vaginatum* is the dominant plant; and, wherever this plant occurs in quantity, the depth of the peat is being added to year by year and denudation of the peat is not táking place. At the present time, *Eriophorum vaginatum* probably forms peat at a more rapid rate and over wider stretches of English moorlands than any other plant; and the statement, occasionally met with, that peat formation is a phenomenon of the past and not of the present is incorrect.

Bog-mosses (*Sphagnum* spp.) are even rarer on the cotton-grass moors than on the heather moors, though a contrary opinion has gained credence; and the erroneous view is still met with that the dominance of Sphagnum is a necessary condition of peat formation. As a matter of fact, Sphagnum is invariably absent from the peat of true fens, and is by no means a necessary constituent of the peat of moors. One may walk many miles over the moors of this district without seeing any trace of Sphagnum; and one may examine many sections of the peat of the district without finding any trace of its remains.

The cotton-grass moors are extensive, dreary, and monotonous. *Eriophorum vaginatum* is frequently not merely the dominant but the only vascular plant which occurs. In late summer and early autumn, the dead green hue of the shoots of the cotton-grass is scarcely relieved by any other touch of colour. In late autumn and throughout the winter, the shoots fade to dull red; and the vegetation then presents a most forbidding aspect. A little life is infused into the area in April and May, when the dusky brown florets make their appearance; but only in June, when the pure white fruits of the cotton-grass appear like suspended snow-flakes, is the cotton-grass moor attractive to the eye (see figure 26).

The monotony of the cotton-grass moor is, however, relieved by certain physiographical features to which the vegetation responds. A sandstone escarpment or outcrop causes a decrease

Copyright *W. B. Crump*

Figure 28.

Retrogressive Moor.

The moor exhibits early signs of retrogression, as the peat is being denuded by the stream. The low cliff (2 metres high) of peat on the extreme left is due to denudation. The vegetation on the general plateau is Cotton-grass Moor. In the foreground, the tufts consist of Mat-grass (*Nardus stricta*): behind are the broad leaves of the Cloudberry (*Rubus Chamaemorus*).

of the dominant plant, and an increase of less hydrophilous species, such as the bilberry (*Vaccinium Myrtillus*) and crowberry (*Empetrum nigrum*). The young shoots of both of these are, in early spring, frequently characterized by rich tints of red and brown which enliven an otherwise dreary landscape. An outcrop of shale is marked by a series of springs, around which featureless Juncus swamps (see pages 146 to 150) occur. A steep slope of shale, damp from oozing water, brings about the vivid greenness of grasses, and locally perhaps of a Sphagnum swamp. Footpaths, as in the heather moors, are marked by a line of mat-grass (*Nardus stricta*), which enables the lonely wanderer to pick his way and to avoid the quagmires which lurk between the tufts of the cotton-grass. *Calluna vulgaris, Nardus stricta, Deschampsia flexuosa*, and *Juncus squarrosus* follow the head-streams almost to their sources.

Ferns and horsetails are absent from all parts of the cotton-grass moor: club-mosses are extremely rare; and, whilst species of mosses, liverworts, Algae, lichens, and Fungi occur here and there, few are really common, and none is of general occurrence. The total absence of moorland tarns and valley lakes is not compensated by the artificial reservoirs which are being constructed in the valleys and less frequently on the moors (see figure 36), as the reservoirs harbour no natural aquatic vegetation such as occurs in the Scottish tarns and lochs.

The association of *Eriophorum vaginatum* is also found on the lowland "mosses" of Lancashire and Cheshire; and it would indeed appear to be specially characteristic of the moors of northern England.

Ostenfeld (1908: 947, *et seq.*) does not describe an association of *Eriophorum vaginatum* in the Faeröes, though associations are detailed in which "*Eriophorum*" and "*E. polystachium*" (= *E. angustifolium*) respectively are said to be dominant. Pethybridge and Praeger (1905) do not find an association of *Eriophorum vaginatum* in the northern Wicklow mountains, where, it would appear, associations of *E. angustifolium* and of *Scirpus caespitosus* hold the same zonal relationship to heather moors that the association of *E. vaginatum* does on the Pennines.

The following short list includes all the flowering plants which have been met with, away from streamsides and footpaths, in typical examples of the association of *Eriophorum vaginatum* of the southern Pennines:—

Dominant

Eriophorum vaginatum

Locally sub-dominant

Molinia caerulea
Eriophorum angustifolium

Locally abundant

Empetrum nigrum
Erica Tetralix
Calluna vulgaris
Vaccinium Myrtillus
Scirpus caespitosus
Carex curta

Local or rare

Andromeda Polifolia
Vaccinium Oxycoccus
Pinguicula vulgaris
Agrostis canina
Narthecium ossifragum

TRANSITIONAL ASSOCIATION OF COTTON-GRASS MOOR AND SILICEOUS GRASSLAND

It has been already stated that parts of the siliceous grassland show transitions to the heather moor: such places have a comparatively dry soil. Some wet parts of the siliceous grassland show analogous transitions to the cotton-grass moor. This transitional association is characteristic of wet and stagnant hollows. The mat-grass (*Nardus stricta*) is less abundant here than on typical siliceous grassland; and the moor grass (*Molinia caerulea*) is frequently conspicuous. *Juncus squarrosus* is sometimes very abundant; but in all cases one or other of the two cotton-grasses, generally *E. angustifolium*, is the most prominent plant. Such areas in the Peak District are local in their distribution. Perhaps the best of them occur between Hayfield and Chinley. The association probably represents a stage in the development of cotton-grass moor from siliceous grassland. On the vegetation maps, they are shown by printing the word "peat" on the colour used for siliceous grassland.

Copyright *W. B. Crump*

Figure 29.

Retrogressive Moor.

The stream has cut through the peat into the shales below. The peat was formerly continuous: it is now dissected into numerous isolated patches or "peat hags."

The relationships of siliceous grassland and moorland may be shown in the following table:—

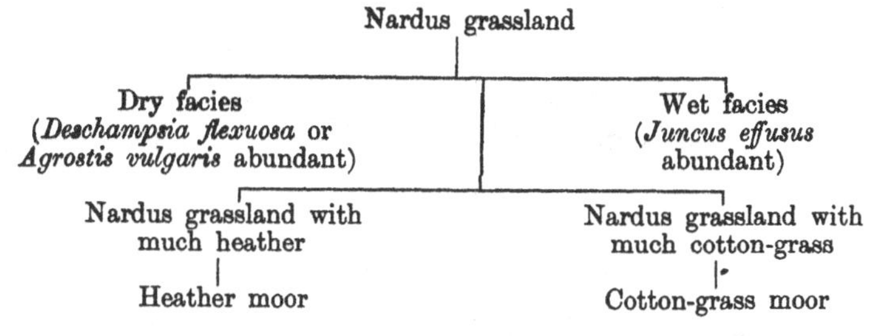

TRANSITIONAL ASSOCIATION OF HEATHER MOOR AND COTTON-GRASS MOOR

Where the boundary between a heather moor and a cotton-grass moor is not marked by an escarpment, there is a wide, level or gently sloping zone in which the heather and the cotton-grass are co-dominant. These transitional areas are marked on the vegetation maps by stippling red dots over the colour used for the cotton-grass moor. A glance at the map will show that such areas are abundant and widespread, especially east of Derwent dale and south-west of Buxton. On many of the latter moors, the cross-leaved heath (*Erica Tetralix*) is very abundant. The majority of the associates of the heather moor are absent; and because of this fact, and because of the deep wet peat which occurs, the transitional moors have, on the whole, more in common with the cotton-grass moors than with the heather moors. Hence the ground colour chosen to indicate the transitional moors on the vegetation maps is that used for the cotton-grass moors. The usual composition of the transitional moors (see figure 27) is given in the following list:—

Sub-dominant

Eriophorum vaginatum — Calluna vulgaris
Erica Tetralix

Locally abundant

Vaccinium Myrtillus — Eriophorum angustifolium
Rubus Chamaemorus — Scirpus caespitosus
Molinia caerulea

Occasional or rare

Deschampsia flexuosa	Narthecium ossifragum
Nardus stricta	Pinguicula vulgaris

Moors of this transitional type have been described as occurring on all portions of the Pennines which have been investigated.

RETROGRESSIVE MOORS

The bilberry (*Vaccinium Myrtillus*), in addition to its being the chief plant on the screes, edges, and ridges of the sandstone rocks, also becomes exceedingly prominent on the peat which is in process of denudation on many of the highest watersheds and plateaux.

Woodhead (1906: 351) appears to think that the occurrence of the bilberry may perhaps always indicate the site of former forest; and he quotes Früh and Schröter (1904) almost to the same effect. This opinion, however, is scarcely applicable to the Pennines where the bilberry occurs abundantly in situations which do not resemble its Alpine habitats.

Whilst the peat of the closed association of *Eriophorum vaginatum* is still increasing in thickness at a comparatively rapid rate, and that of the closed associations of heather and bilberry is also increasing though much more slowly, the peat on the most elevated portions of the moors is gradually being washed away. This process of physical denudation represents a stage through which, it would appear, all peat moors, if left to themselves, must eventually pass. Following Cajander (1904: 1 and 35—37), the associations thus formed are termed retrogressive [" regressive "] associations.

In the Peak District, the process of retrogression in the cotton-grass moors is apparently initiated by the cutting back of streams at their sources. For example, the streams on the Peak are shown, on the revised Ordnance survey maps (1870—1880), to be nearly three-quarters of a mile (1·2 km.) longer than they were when the Peak was originally surveyed in 1830; and they are now a quarter of a mile (0·4 km.) longer than they are shown to be on the revised maps of 1879. The channels formed by the streams which have thus eaten their

Copyright *A. Wilson*

Figure 30.

Retrogressive Moor.

A later stage of retrogression of the Cotton-grass Moor. Deep peat, capped with Cotton-grass (*Eriophorum vaginatum*), etc. Broken Millstone Grit in the bed of the stream.

way back have their banks (see figure 29) fringed with sloping banks of bare peat. In times of drought, the bed of these streams contains very little water which may temporarily disappear; but after heavy rain-storms, the stream is a rapid torrent of brown, peaty water. Every storm results in quantities of peat being carried away, in the stream winning its way further back into the peat, and in the channels becoming wider and deeper. Numerous tributary streams also are formed in course of time; and eventually the network of peaty channels at the head coalesces with a similar system belonging to the stream which flows down the opposite hill-side. The peat-moor which formerly was the gathering ground of both rivers, is thus divided up into detached masses of peat, locally known as "peat-hags" (figure 31); and the final disappearance of even these is merely a matter of time.

It is obvious that this process results in a drying up of the peat of the original cotton-grass moor; and it is most interesting to trace a series of degradation changes of the now decaying peat moor. The first change of importance of the vegetation appears to be the dying out of the more hydrophilous species, such as *Eriophorum vaginatum* and *E. angustifolium*, and the increase, on the summits of the peaty "islands" or "peat-hags," of plants, such as *Vaccinium Myrtillus* and *Empetrum nigrum*, which can tolerate the new and drier soil conditions.

The composition of the upper layers of the peat of these retrogressive moors has, during the course of the present investigation, been carefully examined; and it has been found that the peat consists in its upper layers almost wholly of the remains of Eriophorum. The succession therefore of cotton-grass moor to the series of retrogressive moors here being described, is established beyond doubt.

The Peak of Derbyshire

As the Peak of Derbyshire is covered by retrogressive moorland, a short description of this the most important topographical feature of the district will not be out of place at this juncture (cf. figure 32).

The Peak is a plateau of Kinderscout sandstone varying in height from about 1750 feet (533 m.) at its eastern extremity to 2088 feet (636 m.) at Soldier's Lump. The latter is the highest altitude attained by any Pennine summit south of the Great Whernside group. The Peak occupies an area of about three square miles (nearly 80 ares). It is clad throughout its entire length and breadth with peat which is about twelve feet (363 cm.) deep on the average. The peat is dissected by very numerous stream channels, formed in the manner just indicated. The summits of the resulting "peat-hags" are, on the whole, dominated by the bilberry (*Vaccinium Myrtillus*); but the crowberry (*Empetrum nigrum*) and the cloudberry (*Rubus Chamaemorus*) are locally very abundant, forming plant societies. Here and there extensive patches of bare peat occur.

The bulk of the peat, all in fact except the lowest layer, is composed of the remains of the cotton-grass. The lowest layer is black, very much compressed, and very deficient in air. When wet or damp, this layer is slippery, like wet soap, to the touch: when dry, it is sometimes brittle and shiny, not altogether unlike Whitby jet. Such peat, which is typical of the highest peat moors, is quite structureless; and one can only speculate as to the plants of which it is the remains. On the sandstone, underneath the peat, there is a thin layer of brittle, reddish-brown ferruginous "pan" ("Raseneisenstein"): on the shale, however, true "pan" ("Ortstein") occurs below the peat (cf. Tansley, 1911: 103). Remains of trees have not been found on the plateau of the Peak, but only on its slopes, where remains of birch were noted in a gully at an elevation of about 1800 feet (549 m.).

The Peak is not an imposing mountain. Viewed from the east or south, only its grassy slopes can be seen. From Ashop dale, on the north, the Edge, as it is called, of sandstone rock stands out rather boldly. From the west, the steep and rocky slopes of Kinderscout provide a wild and picturesque landscape. This view is especially fine, in the spring when the young red and green shoots of the bilberry, and in the autumn when the richer brown and golden colours of the dying fronds of the bracken contrast with the sombre green of the heather and crowberry and the forbidding blackness of the precipices and large and loosely scattered boulders. Only on the Glossop and Sheffield high road, at its highest elevation four miles out of Glossop, may a general view of the summit be obtained; and

 A. Wilson

Figure 31.

Retrogressive Moor.

A still later stage of retrogression of the Cotton-grass Moor. Peat-hags with much bare peat, capped here and there with Bilberry (*Vaccinium Myrtillus*).

this, the only view obtainable of the summit as a whole, is dull and uninspiring in the extreme.

From the standpoint of floristic botany, the Peak is uninteresting; still, it is of interest to the ecologist as it provides many problems in the succession of plant associations.

The following plants occur on the plateau or uppermost slopes of the Peak:—

Dominant species

Vaccinium Myrtillus

Locally sub-dominant species

Rubus Chamaemorus
Empetrum nigrum
Vaccinium Vitis-idaea[1]

Occasional or locally abundant species

Potentilla erecta[1]
Galium saxatile[1]
Calluna vulgaris[1]
Molinia caerulea[1]
Nardus stricta[1]
Eriophorum angustifolium
Juncus squarrosus[1]
J. effusus[2]

Local or rare species

"Lycopodium spp."
Blechnum spicant[1]
Nephrodium dilatatum[1]
"Arctostaphylos Uva-ursi"
Erica Tetralix[1]
Festuca ovina[1]
Deschampsia flexuosa[1]
Scirpus caespitosus[1]
Eriophorum vaginatum
"Carex dioica"
C. curta
"Listera cordata"

BARE PEAT

As the process of peat-denudation proceeds, the members of this retrogressive plant association gradually succumb to the changing conditions, until the "peat-hags" become almost or quite bare of plants. At this stage, there is nothing to hold the peat together; and it is washed and whirled about by every rainstorm, and by the waters of melting snow. Such bare peaty summits are of great extent on Black Hill, Holme Moss (cf. Smith and Moss, 1903: 382), on parts of the plateau of the Peak, and occur to a greater or less extent on most of the exposed summits of the Pennines. Almost the only plants

[1] Chiefly at the edges of the Peak.
[2] Confined to stream sides.

to be found on such extremely decadent moors are a few straggling and miserably developed specimens of *Eriophorum angustifolium*. The words "Bare peat," printed on the vegetation maps here and there, roughly indicate the spots where the more extensive of the tracts occur such as are here described.

As has been stated, an examination of the peat deposits underlying the retrogressive peat-moors here described proves that it is composed almost wholly of the remains of cotton-grass; and the living Vaccinium and Empetrum which crown the "peat-hags" rest unconformably—as the geologists would say—on strata of cotton-grass peat. Hence the conclusion may be safely drawn that the retrogressive phases characteristic of the highest Pennine plateaux are very recent in origin, and, in all probability, have been initiated during the last few centuries. The process is still at work, and is likely to become more and more pronounced as time goes on.

The decadent condition of many of the summits of the Pennine peat moors make it an easy task to determine that the ancient Pennine forest did not, at any period, spread over the highest summits; as, although the base of the peat is very frequently exposed, remains of timber have nowhere been found on the highest summits. In addition to the examination of the peat which is being denuded on the high summits, several sections have also been cut with the spade, and with the same negative results. The retrogressive changes appear, in many cases, to be spreading downwards into the lower cotton-grass moors; but many of the latter show no signs of degeneracy as yet. The heather moors also are generally speaking in a state of stability at the present time.

On the vegetation maps, the more pronounced of the retrogressive moors are indicated by the hatching of red lines on the Eriophorum colour. It is reasonable to use this ground colour as the evidence shows the moors to have been Eriophorum moors until quite recent times, and the retrogressive changes are still in operation. Owing to the comparative inaccessibility of these moors, the absence of landmarks upon them, the absence of contour lines on the six-inch Ordnance maps, and the impermanent nature of the plant association, the boundaries of these retrogressive moors were difficult and in many cases impossible to fix with accuracy.

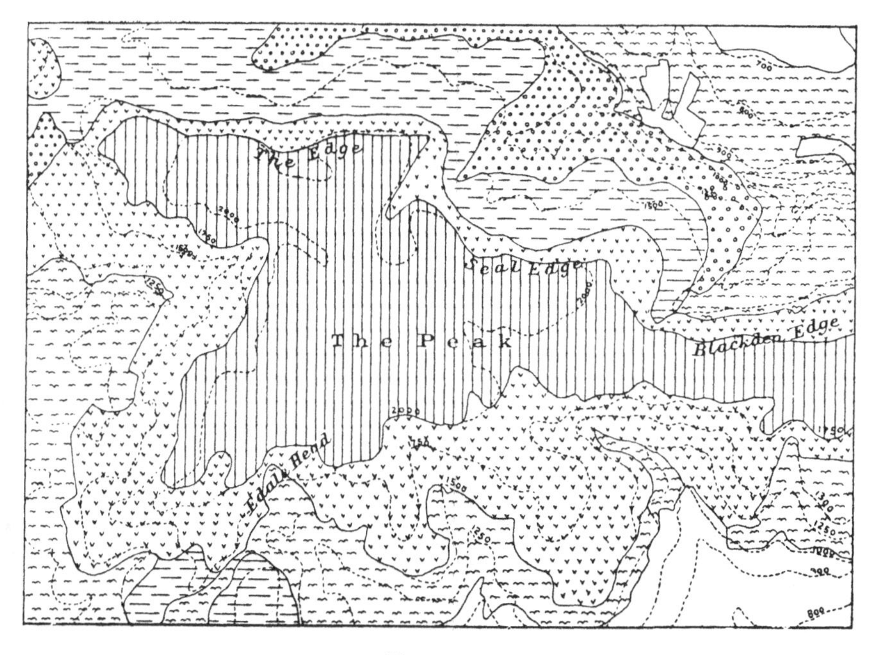

Figure 32.

Map of the Vegetation of the Peak of Derbyshire.

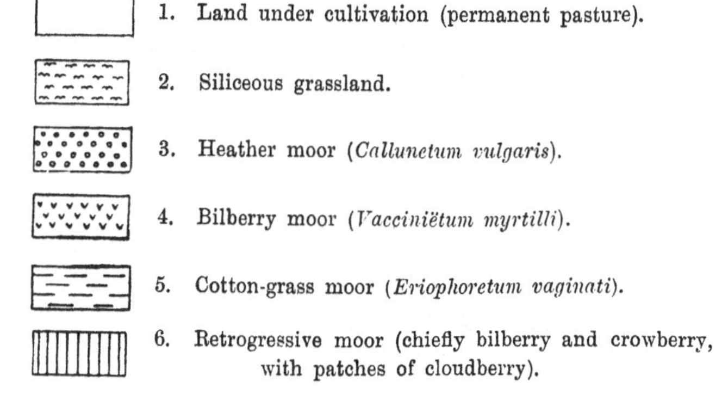

The rocks consist entirely of sandstone and shales (Pendlesides and Millstone Grits).

SUB-ALPINE GRASSLAND

In the end, the retrogressive changes outlined above result in the complete disappearance of the peat; and on the surface thus laid bare, a new set of species begins to invade. In this invasion the ordinary inhabitants of the surrounding peat-moors can take no part; and the successful invaders are the more hardy members of the Nardus grassland. As has been stated (see page 185), such plants follow the streams of the peat-moors almost to their sources; and hence they are the plants which one would expect to be the first to establish themselves in the newly formed habitat. The summit of Bleaklow Hill, four miles north of the Peak, is tenanted by an open plant association which has almost certainly originated in the way just described. On the summit of Great Whernside and other hills of the mid-Pennines, Smith and Rankin describe an association which seems to be capable of a similar interpretation: "the summit-ridge from the edge of the peat-bog upwards is rocky, with a scanty soil which supports a meagre vegetation consisting of grasses. Here and there are patches or islands of peat" (Smith and Rankin, 1903: 154).

The following species were observed on the summit of Bleaklow Hill:—

Nardus stricta
Deschampsia flexuosa
Festuca ovina
Agrostis vulgaris
Juncus squarrosus
Luzula erecta
Rumex Acetosella
Potentilla erecta
(= P. Tormentilla)
Calluna vulgaris
Vaccinium Myrtillus
Galium saxatile

On Great Whernside, Smith and Rankin (*loc. cit.*) record *Festuca ovina* forma *vivipara* and *Poa alpina*; and it is not impossible that a careful search on Bleaklow Hill would reveal these plants, although they have not yet been recorded for Derbyshire.

The case of this sub-Alpine Pasture illustrates the important principle that a succession of plant associations, once initiated, may lead from one plant formation to another (cf. Moss, 1910 *b*: 37). In the case under discussion, the retrogressive succession began in the closed cotton-grass association, continued through the retrogressive associations of bilberry, and, as regards the moorland formation, ended with bare peat. This open phase of a retrogressive series of changes terminates a formation, just as, in a progressive succession an open association is the starting point of a formation. In the case under consideration, however, the succession has not ended with the terminal association of the moorland formation, but has been continued by an open grassland association. What the future of this succession will be is a matter of speculation; but one may easily imagine, assuming climatic conditions to remain unaltered, that the future of this succession will show a closed association of grassland, of mixed grassland and heath, and later, in the still more distant future, of some moorland association.

The matter of this succession has been discussed here at some length in order to show that it is possible to account for the changes which the moorland vegetation has in comparatively recent times undergone, on other than climatic grounds.

Zonation of the Moorland and Grassland Associations

The zonation of the moorland and grassland associations of the non-calcareous summits and slopes of the southern Pennines is exceedingly well marked, as a study of the vegetation maps will themselves testify. In a general way, the zonation of plant associations on any mountain illustrates the effects of altitude on vegetation; but these effects are usually modified to some extent by some local conditions. In this district, the local conditions which compete with altitude in modifying the zonation are chiefly physiographical in character. However, the combined effects may be stated in general terms, if one speaks of the broad outlines of the vegetation and ignores details.

Those eminences which are capped by a fairly flat plateau are characterized by summits which are covered with retrogressive moorland associations (figure 33, *a*). These retrogressive associations are surrounded by rocky escarpments covered with stable associations of bilberry (figure 33, *b*). More pointed eminences are capped by a stable bilberry moor (figure 34, *b*). The bilberry moors are surrounded by moderately elevated, shelving plateaux of only moderate steepness; and on these plateaux a broad zone of cotton-grass moors (figures 33 and

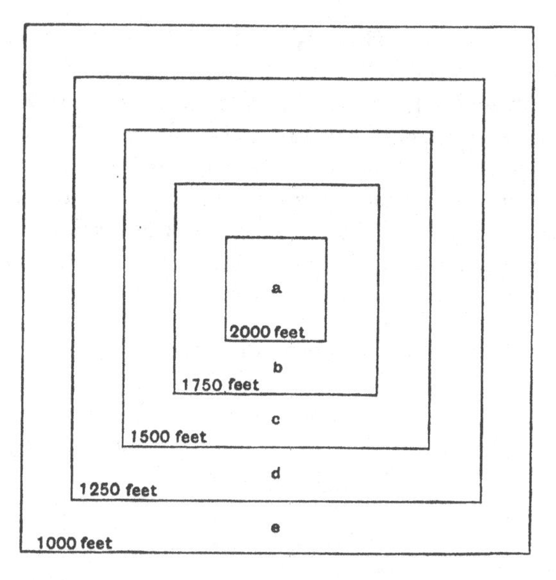

Figure 33. Diagram of the vegetation of a flat-topped eminence reaching an altitude of about 2000 feet (610 m.), *e.g.*, the Peak.

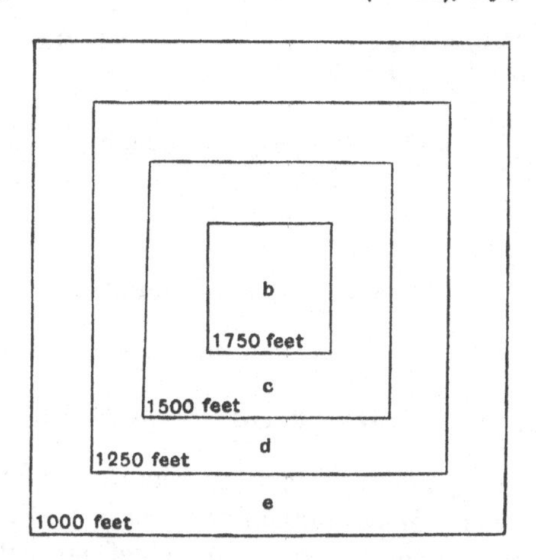

Figure 34. Diagram of the vegetation of a more pointed eminence reaching an altitude of about 1750 feet (533 m.), *e.g.*, Mill Hill, near Hayfield.

a. Retrogressive moorland associations.
b. Bilberry moors.
c. Cotton-grass moors.
d. Either heather moors or siliceous grassland.
e. Upland cultivation.

34, *c*) is developed. The plateaux of the cotton-grass moor either descend gradually into the zone of upland cultivation (figures 33 and 34, *e*), in which case these two zones are separated by a zone of heather moor (figures 33 and 34, *d*); or they are terminated abruptly by steep shaly slopes, in which case the zones of cotton-grass moor and upland cultivation are separated by a zone of Nardus grassland (figures 33 and 34, *d*).

The relationships of the plant formation of the siliceous soils and that of the moors may be seen in the following table:—

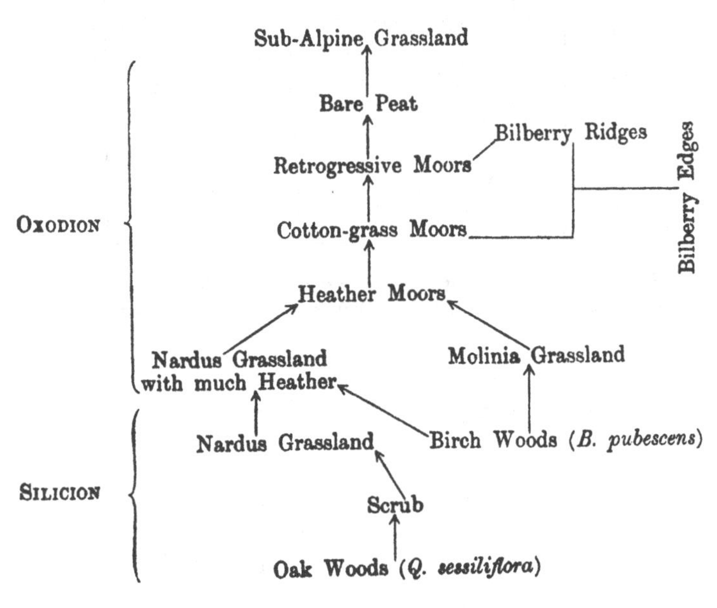

List of Species of the Moor Formation

The following species occur on the moor formation (cf. Ostenfeld, 1908: 947 et 956) of the southern Pennines; and their relative frequency in the three chief associations is also indicated. The plants preceded by an obelisk have not been recorded from the Peak District; but they occur a few miles to the north.

	Heather moor	Bilberry moor	Cotton-grass moor
Sphagnum spp.	la	—	r
Polytrichum spp.	l	l	—
Hypnum spp.	l	—	o
Lycopodium spp.	vr	vr	—
† Selaginella selaginoides	—	vr	—
Blechnum spicant	o	r to o	—
Pteris aquilina	r to la	r to o	r
Nephrodium aristatum	r	r	—
Salix repens	vr	—	—
Betula pubescens (dwarfed)	r	r	—
Quercus sessiliflora (dwarfed)	r	r	—
Rumex Acetosella	l	r to o	—
R. Flammula	l	—	—
"Drosera anglica"	vr	—	—
D. rotundifolia	r	—	—
Potentilla erecta (=P. Tormentilla)	o	r to o	—
"P. palustris"	vr	—	—
Rubus Chamaemorus	—	r, la	r, ls
Crataegus Oxycantha (=C. monogyna) (dwarfed)	r	r	—
Pyrus Aucuparia (dwarfed)	r	r	—
Genista anglica	r	—	—
Ulex Gallii	la	—	—
Lathyrus montanus	r	—	—
Polygala serpyllacea	r	—	—
Empetrum nigrum	la	la	la
Ilex Aquifolium (dwarfed)	r	—	—
Viola palustris	l	—	—
Hydrocotyle vulgaris	l	—	—
† Pyrola media	vr	—	—
Andromeda Polifolia	vr	—	r
Arctostaphylos Uva-ursi	—	la	—
Erica cinerea	o to s	la	—
E. Tetralix	la	—	la
Calluna vulgaris	d	ls	la
var. Erikae	o	—	—
forma incana	vr	—	—
Vaccinium Myrtillus	o to s	d	la
V. Vitis-idaea	la	ls	—
V. Oxycoccus	l	—	l
† Trientalis europaea	vr	vr	—
Melampyrum pratense	r	—	—
Pedicularis sylvatica	l	—	—
"P. palustris"	vr	—	—
Pinguicula vulgaris	r	—	r
Galium saxatile	la	la	—
Cnicus palustris	l	—	—

	Heather moor	Bilberry moor	Cotton-grass moor
Agrostis canina	l	—	l
A. tenuis (=A. vulgaris)	l	--	—
Aira praecox	l	—	—
Deschampsia flexuosa	o to a	a	—
Molinia caerulea	r to o	r	r, ls
Festuca ovina	r to o	—	—
Nardus stricta	r to o	r to o	—
Eriophorum vaginatum	l	l	d
E. angustifolium (=E. polystachiom)	l	—	la
Scirpus caespitosus	o	—	l
"S. pauciflorus"	vr	—	—
"Carex dioica"	vr	vr	—
C. echinata (=C. stellulata)	l	—	—
C. curta	r	--	la
C. Goodenowii	la	—	—
var. juncella	r	—	—
C. flacca (=C. glauca)	l	—	—
C. pilulifera	r	r	—
C. panicea	l	—	—
C. binervis	r	r	—
C. flava	la	—	—
forma minor	l	—	—
Luzula erecta	o	—	—
forma congesta	r to o	—	—
Juncus squarrosus	la	—	—
J. effusus	l	—	—
J. articulatus (=J. acutiflorus)	l	—	—
Narthecium ossifragum	l	—	r
Orchis ericetorum	r	—	—
"Listera cordata"	?ext.	—	—

CHAPTER VIII

CULTIVATED LAND: CULTURE ASSOCIATIONS

Origin of the cultivated land. Nature of the cultivated land. Permanent pasture. The arable land. Plantations. Afforestation. Utilization of the peat-moors.

ORIGIN OF THE CULTIVATED LAND

THE whole of the land now fenced and under cultivation was, of course, originally occupied by spontaneous plant associations. Of these, those that once covered the land now cultivated were in all probability of the nature of woodland in prehistoric and even in early historic times.

Most of the land below about 900 feet (274 m.) has been cultivated for many centuries; but there is historical evidence which shows that, during the last century and a half, considerable intakes at altitudes up to about 1500 feet (457 m.) have taken place. Whilst the process of reclamation is, to a slight extent, still proceeding, the modern attempts in this direction are of a local and intermittent character. These attempts are nearly all made at the expense of grassland or the lower fringe of the moorland.

NATURE OF THE CULTIVATED LAND

Almost the whole of the cultivated land of this district is laid down to grass, and is termed by English agriculturists "permanent pasture," as it is nowadays never ploughed. Ploughed or arable land is, on the whole, of rather uncommon occurrence.

The cultivated land is separated by fences constructed of either sandstone or limestone. The sandstone walls ultimately weather to an almost black hue, whilst the limestone walls remain white; and the presence of black or white stone fences is a convenient indication as to whether one is in the area of sandstone or of limestone respectively.

Hedgerows in the Pennine district are rare, and only occur where the shales are of great superficial interest, as, for example, near the confluence of the rivers Noe and Derwent.

At its upper limit, the permanent pasture frequently abuts on the uncultivated grassland. A distinction is made on the Ordnance maps between "land under cultivation" and "land not under cultivation"; but, at and near the upper limits of cultivation, the boundaries shown on the Ordnance maps are not always reliable. A comparison of the boundary line between cultivated and uncultivated land as shown respectively on the Ordnance maps and on the accompanying vegetation maps will reveal rather considerable discrepancies.

I am unaware of the principles used by the Ordnance surveyors in making this distinction. In the present vegetation survey, the plan has been to make lists of the species of the difficult tracts, and compare the lists thus made with lists of tracts which are indubitably uncultivated or cultivated, as the case may be. There are, without doubt, many areas with regard to which there may be differences of opinion as to whether or not they should be mapped as land under cultivation; but this does not explain all the details of the mapping of the Ordnance surveyors, who, indeed, are sometimes very inconsistent even on the same "six-inch" quarter-sheet.

On the accompanying vegetation maps, the grassland not considered to be cultivated, although it may be more or less grazed, is coloured as siliceous grassland when the flora contains many heath-loving or humus-loving species, and coloured as calcareous grassland when there are many lime-loving species present. These two associations or groups of associations have been discussed in the chapter on grasslands. Some of the enclosed fields have apparently once been cultivated and have been allowed to become derelict; and such areas, by the invasion of plants from the uncultivated land, gradually approach in character to the neighbouring subspontaneous or spontaneous associations. However, up to about 1250 feet (379 m.) the cultivated fields may generally be kept in good condition without much difficulty; and one frequently sees, even at the very edge of a Calluna moor, bright green permanent pasture which shows no tendency to revert to its original state (cf. figure 24).

During the course of this survey, the process of reclamation has been observed in a few cases. The plan adopted was as follows. The original vegetation, whether heather (*Calluna vulgaris*) or grasses (*Nardus stricta*, etc.) was first burned, and then cleared of large stones. The land was afterwards ploughed and limed, and finally planted with oats. The field sometimes remained a patch of arable land; but more frequently, grasses were sown in the second or third years, and the land kept down to permanent pasture. In some cases, but by no means all, the land was also drained by means of trenches and agricultural drain pipes. Where the original land was covered with shallow peat, the peat was flaked off before the land was ploughed. Deep peat on these uplands is practically never reclaimed; and hence the soil of the cultivated uplands is rarely black, though it may be of a very dark brown colour owing to its high humus-content.

Even on the upland tracts which are now almost wholly cultivated, it is frequently possible to form definite and accurate ideas regarding the nature of the natural plant associations which were formerly characteristic of the places in question; for some of the indigenous species often linger in some not wholly unsuitable localities. Such places are the grassy or heathy banks and sides of the roads and lanes which are not much frequented, quarries, gravel pits, refuse heaps of old mines, old hedgerows, hedgebanks, hedgebottoms, and the banks of streams. Although such localities usually contain a mixture of indigenous and alien plants, it is seldom impossible to decide to which of these categories a given species belongs.

The farms of the district are of small size, and rarely consist of more than forty or fifty acres (1620 or 2025 ares). It is said by some of the farmers that rather more land was under the plough some forty years ago; but the district as a whole has never been important in the matter of corn growing. Before the days of cheap flour, probably each farm produced its own oatmeal at least; but there is no evidence to show that any crop of the district was ever of more than domestic importance.

PERMANENT PASTURE

The permanent pasture, although nowadays never ploughed, is an artificial plant association or group of artificial associations. Human influence is seen in three ways. First, most of the permanent pasture has been ploughed at least once, and in many cases has been sown with grass seeds, and the original plant associations have therefore been destroyed: secondly, it is more or less regularly manured by the occupying farmers: and thirdly, it is always grazed over by cattle, horses, or sheep.

The manuring and grazing effectually prevent many of the aboriginal species re-migrating into the area. However, when permanent pasture becomes neglected and derelict, these species tend to enter the area and the introduced species tend to die off. Ultimately, the derelict pastures are indistinguishable from the uncultivated grassland; and they are therefore so coloured on the accompanying vegetation maps.

Many of the cultivated grass fields are utilized solely for grazing purposes. In a considerable number of cases, however, the cattle are kept out of the fields after April; and the grass is allowed to grow long, when it is cut for hay. In this district haymaking usually begins about the end of June and continues until the middle or end of August, or into September if the season is unpropitious. By way of a phenological comparison, it may be stated that in eastern Somerset, haymaking begins at the end of May or the beginning of June. After the hay has been cut, the cattle are again turned into the fields; hence, it is scarcely possible to make any distinction on the maps between "meadows" and "pastures."

The most abundant grasses of the hay-fields are:—

Agropyrum repens	a	Dactylis glomerata	a
Alopecurus pratensis	a	Festuca rubra	a
Anthoxanthum odoratum	a	Lolium perenne	la
Phleum pratense	l	Poa pratensis	a
Bromus mollis	la	P. trivialis	l, ?r
Cynosurus cristatus	la	Trisetum flavescens	o

Of the above grasses, Cynosurus is characteristic of the drier and poorer soils, and Trisetum of the damper and richer soils.

The plants in the following list are mostly counted as "weeds" by the farmer. They are most abundant in the fields bordering on the uncultivated land, which are less frequently and less regularly manured than the fields at lower levels. Most of the species are really indigenous to the district, and would perish if the fields in which they occur were manured more systematically. The list contains most of the species of the more upland permanent pastures on the sandstones and shales, although, owing to the manuring, there is no great difference between the permanent pastures of the sandstones and shales and those of the limestones:—

In drier pastures		In damper pastures	
Ophioglossum vulgatum	r, la	*Rumex alpinus	vr
Pteris aquilina	r, la	Stellaria graminea	o
Rumex Acetosella	o, la	Ranunculus repens	la
Polygonum Bistorta	o, ls	R. bulbosus	o, la
Cerastium vulgatum	o	R. acris	a
Potentilla erecta	o	Saxifraga granulata	vr
P. procumbens	la	Alchemilla pratensis	o to a
Lotus corniculatus	o	Sanguisorba officinalis	la
Trifolium medium	l	Trifolium repens	o to a
Lathyrus montanus	l	T. pratense	o
Hypericum pulchrum	r	Anthriscus sylvestris	l
Viola lutea	r, la	Heracleum Sphondylium	o
Pimpinella Saxifraga	r to o	Conopodium majus	o
Veronica officinalis	la	Prunella vulgaris	o
Euphrasia officinalis	la	Ajuga reptans	o
Rhinanthus Crista-galli	la	Veronica Chamaedrys	o
Plantago lanceolata	a	Achillaea Ptarmica	l
Galium saxatile	la	A. Millefolium	a
Campanula rotundifolia	r to o	Bellis perennis	a
Centaurea nigra	o to a	Senecio Jacobaea	la
Chrysanthemum Leucanthemum	o to a	Hypochaeris radicata	o
Leontodon hispidum	o	Leontodon autumnale	a
Crepis virens	o	Taraxacum officinale	a
Hieracium Pilosella	r to a	Holcus lanatus	o
Agrostis vulgaris	la	Deschampsia caespitosa	l
Briza media	l	Carex ovalis	l
Festuca ovina	r, la	*Narcissus Pseudo-narcissus	r, la
Luzula campestris	la	*Crocus nudiflorus	la

Near the upper limits of cultivation, the manuring often consists of dressings of lime or of farmyard manure; and it is only as the lowlands are approached that chemical manuring is freely utilized.

The Arable Land

In previous accounts of the vegetation of districts in Great Britain, it has been customary to give a table, taken from the official Agricultural Returns, showing the amount of the various types of the agricultural land of the county in which the district is situated. In the case of the Peak District, such a course is undesirable, as the land consists of portions of five counties and is very far from being typical of any one of them. Of the various English counties, the Peak District most nearly resembles Westmorland (cf. Lewis, 1904 *a*: 316) in its high percentage of uncultivated land, and in its low percentage of arable land, especially of land under wheat.

At the present time, it is possible to sub-divide the cultivated land of the British Isles into three zones (cf. Moss, 1907 *a* or *b*: 21, 59, 66). The lowest of these zones, occurring as a rule below fifty feet (15 m.) above sea level, consists of alluvial land: in the west of England (Moss, 1907 *a* or *b*: 21), this alluvial zone of cultivation is nearly all under permanent pasture; but in East Anglia, it is nearly all under arable cultivation, with wheat entering into the rotation. The intermediate zone, situated as a rule below six or seven hundred feet (183 or 213 m.) above sea level, consists largely of permanent pasture in the west and north of England: it shows a higher proportion of arable land, with wheat entering into the rotation, in the Midlands and in the south of England; and it consists very largely of arable land, with wheat, in East Anglia. The uppermost zone, situated as a rule above six or seven hundred feet above sea-level, consists largely and in many localities almost wholly of permanent pasture; and in the arable land that actually occurs, wheat does not enter into the rotation, or, if so, it is a crop of a precarious nature.

In the Peak District, no zone of alluvial cultivation occurs; but it has been found possible to show on the map the intermediate (or wheat) zone and the uppermost (or no-wheat) zone: a transitional zone is indicated on the map by stippling. There can be no doubt that the dividing line between the wheat and the no-wheat zones is drawn on the vegetation maps with considerable accuracy. In several cases, upland fields of wheat have been closely observed during a series of years in order to determine the effect of good and bad seasons on the ripening of the grain at or near the upper limit of wheat cultivation. For example, several wheat fields in Derwent dale and in the Hope valley were uncut on October 30th, 1906; and, after such a date, wintry types of weather may, in this locality, be expected at any time, and actually came in early November in 1910. It is clear therefore that the wheat fields of the locality in question represent the upper climatic limit of wheat in the Peak District. Generally it is claimed for the vegetation maps of Great Britain that they represent the limits of wheat cultivation more accurately than has been done on any other maps in any country or at any time; and, from this point of view alone, the maps are of great value.

The upper limits of wheat cultivation in the southern Pennines vary somewhat on the different soils. On the eastern plateaux of the Coal-measures, wheat is usually grown up to 700 feet (213 m.), rarely up to 900 feet (274 m.), and most rarely up to 1000 feet (305 m.). On the Pendleside (or Yoredale) shales and river gravels in the Hope and Derwent valleys, wheat is usually grown up to 600 feet (183 m.) and rarely up to 850 feet (259 m.). On the Millstone grit, wheat is rare generally, and has not actually been observed higher than 500 feet (152 m.). On the Mountain Limestone, not a single case of wheat cultivation has been observed. On the other hand, oats (*Avena*) is not infrequently grown on all the soils up to 1250 feet (381 m.) and more rarely up to 1350 feet (411 m.). Oats are much more commonly grown on the limestones than on the sandstones. Cereal crops, other than wheat and oats, are quite rare. Barley (*Hordeum*) is rarely grown, and rye (*Lolium*) scarcely at all. In the no-wheat zone, the rotation is of a very primitive character, oats being often grown several years in succession, or, more rarely, in a twofold rotation with roots, usually turnips (*Brassica*). In the wheat zone, the usual fourfold rotation—wheat, roots, oats, clover (*Trifolium*)—is frequently followed.

From the above facts, it will be seen that wheat is cultivated up to its local climatic limit, but that this varies on the different soils, being highest on the shales of the Coal-measures and Pendlesides and lowest on the Millstone Grit and the

Limestone. Oats are apparently less responsive to soil factors, and are grown on all classes of soils. On any given soil, the dividing line between the wheat and the no-wheat zone represents a limit determined by climatic conditions; and this limit varies on the different soils.

The prevailing views with regard to the climatic factors determining the successful limits of wheat cultivation are given in the paper on the vegetation of the Leeds and Halifax district (Smith and Moss, 1903: 395—8). It is there stated, chiefly on the authority of Buchan (1862), that wheat requires an average summer temperature of at least 56° F. (13° C.) and a rainfall of not more than about 33 inches (84 cm.). Judging by the present distribution of wheat cultivation in Somerset and in the present district, it would appear that whilst the above figures are approximately correct as regards temperature, the rainfall figures are too low by about ten inches (25 cm.); for wheat regularly ripens in Somerset (see Moss, 1907 *a* or *b*) and in the Peak District, where the mean annual rainfall is over 40 inches (102 cm.).

Previous British vegetation maps have indicated the cultivated land by various tints of yellow, irrespective of the proportion of arable land to permanent pasture. On future maps, it is proposed to make some distinction between cultivated land with a high proportion of permanent pasture, as in the Peak District, and cultivated land with a low proportion of permanent pasture, as in East Anglia; and the maps which accompany the present volume are coloured on this plan.

It has previously been pointed out (Smith and Moss, 1903: 399; Moss, 1907 *a*: 61) that the limit of wheat cultivation corresponds roughly with the limits of a number of weeds and aquatic plants. The following is a list of weeds which have been noted in the arable fields of the district; but it should be borne in mind that, as there is but little typical wheat land in the Peak District, the weeds of the wheat zone are, with regard to the district as a whole, either very local or very rare:—

	Wheat zone	Wheat and no-Wheat zone
Equisetum arvense	la	la
Urtica dioica	o	o
U. urens	o	—
Rumex Acetosa	o	o
R. Acetosella	o	o
R. obtusifolius	o	o
Polygonum Convolvulus	o	—
P. aviculare (agg.)	a	a
P. rurivagum	r	—
P. Persicaria	o	o
P. lapathifolium	r	—
Chenopodium album	o	—
Atriplex patula	a	o
var. angustissima	a	a
Silene Cucubalus	r	—
Lychnis alba	r	—
L. Githago	r	—
Stellaria media	a	a
Arenaria serpyllifolia	r	—
Spergula arvensis	a	a
Ranunculus arvensis	r	—
Papaver Rhoeas	r to o	—
P. dubium	r	—
Fumaria officinalis	r	—
F. pallidiflora	r	—
Cardamine hirsuta	la	—
Brassica nigra	r	—
B. arvensis	a	o
B. alba	r	—
Capsella Bursa-pastoris	a	a
Raphanus Raphanistrum	r	—
Alchemilla arvensis	o	—
Melilotus spp.	r	—
Trifolium spp.	o	o
Geranium molle	o	—
G. pusillum	r	—
G. dissectum	o	—
Euphorbia Peplus	o	o
E. Helioscopia	o	o
E. exigua	o	—
Viola arvensis (agg.)	o	—
V. tricolor (agg.)	—	o, la
Scandix Pecten-Veneris	o	—
Aethusa Cynapium	o	—
Heracleum Sphondylium	o	o
Convolvulus arvensis	o	—
Lithospermum arvense	l	—
Mentha arvensis	o	—

	Wheat zone	Wheat and no-Wheat zone
Stachys arvensis	r	—
Galeopsis angustifolia	r	—
G. versicolor	r	—
G. Tetrahit	o	o
Lamium purpureum	o	—
Veronica polita	r	—
V. agrestis	o	—
V. persica	o	—
V. arvensis	o	—
V. hederaefolia	o	—
Plantago major	o	o
P. lanceolata	o	o
Galium Aparine	a	o
Sherardia arvensis	o	—
Valerianella dentata	o	—
V. olitoria	o	—
Scabiosa arvensis	o	—
Bellis perennis	o	o
Gnaphalium uliginosum	l	l
Achillaea Ptarmica	l	l
A. Millefolium	o	o
Anthemis Cotula	o	—
A. arvensis	r	—
Matricaria inodora	o	o
M. Chamomilla	o	—
Tussilago Farfara	la	la
Senecio vulgaris	a	a
Cnicus arvensis	o	o
C. lanceolatus	o	o
Centaurea Cyanus	r	—
Lapsana communis	o	—
Taraxacum officinale	o	o
Sonchus oleraceus	o	o
S. asper	o	o
S. arvensis	o	—
Holcus lanatus	l	l
Poa annua	a	a
Bromus secalinus	r	—
B. racemosus	r	—
Lolium perenne (agg.)	l	l
Triticum repens	l	l
Juncus bufonius	l	l

Plantations

The natural and semi-natural woods of the district have been described in a previous chapter. It remains to mention the plantations. The latter term is here used to denote purely artificial associations of trees whose ground flora does not comprise shade-loving species.

In this district, plantations occur usually on the site of former grassland, less frequently on a moorland site, and scarcely ever on former arable land.

The trees most commonly planted are the Scots pine (**Pinus sylvestris*), the larch (**Larix decidua* = **L. europaea*), and the beech (**Fagus sylvatica*). Other trees which are locally abundant in the plantations are the black or Austrian pine (**P. nigricans* = **P. austriaca*), the spruce fir or Norway spruce (**Picea excelsa*), the oak (**Quercus Robur* = **Q. pedunculata*), and the sycamore (**Acer Pseudoplatanus*).

On the vegetation maps, the plantations which consist chiefly of coniferous trees are distinguished from those which consist chiefly of dicotyledonous trees by means of special colours; and the most abundant trees are indicated by letters, thus: **Pinus sylvestris* (p), **Larix decidua* (=**L. europaea*) (L), **Fagus sylvatica* (F), mixed conifers (C), mixed dicotyledonous trees (D), or mixed coniferous and dicotyledonous trees (M). In the case of plantations consisting of an approximately equal mixture of deciduous and coniferous species, the fact is indicated on the maps by stippling.

The pine plantations are more numerous on the non-calcareous than on the calcareous soils: larch and beech plantations occur indifferently on either soil. In damp situations, the larch is commonly attacked by canker (*Dasyscypha calycina* = *Peziza Wilkommii*). The beech grows well in the district, more especially perhaps on the limestones; but nowhere on the Pennines does the tree appear to rejuvenate from self-sown seed. Henry (1907: 100) states that the beech is native in this district; but that is not the usual view. The place-name Buxton which Henry infers means "beechtown," is capable of a very different derivation. Lees (1888) says that the beech is "possibly native on the Permian" or Magnesian Limestone of

Yorkshire. It is curious therefore that Linton (1903) should not record the tree from a single station on the Derbyshire continuation of the Permian Limestone of Yorkshire, although one would think it must certainly occur here. Although the matter is a difficult one to settle, the balance of evidence seems to be against the view that the beech is native in Britain so far north as Derbyshire. The tree, however, is indigenous in the south of England, where it forms beech woods (see Moss, Rankin, and Tansley, 1910), especially on the escarpments of the Chalk and on the Greensand.

Many of the plantations are small; and of these only those in sheltered situations are successful. Small plantations in exposed situations are often ruined by the severe and cold winds of the hills; and derelict plantations are far too common on the Pennines (see figure 35).

On the other hand, the larger plantations are, on the whole, in a prosperous condition, especially those in the valley of the Derwent, *e.g.*, the large one north-west of Strines reservoir, and in the Goyt valley, *e.g.*, the still larger one to the south-west of Taxal. The one near Taxal is said to be the most extensive plantation in Cheshire, and to occupy not less than a thousand acres. It was begun about the year 1796—8 (see Holland, 1808: 10). The southern portion of this plantation is composed almost wholly of beech (*Fagus sylvatica*) planted on soil containing sour humus. At the present time, the chief ground species under the beeches is *Deschampsia flexuosa*, but all the commoner plants of the natural heath pasture occur. There are very few other trees or shrubs; but the beeches are vigorous and healthy. This portion of the plantation is wholly below 1250 feet (379 m.).

The more elevated portion of the plantation is composed principally of mixed conifers; and, whilst a large proportion of it is flourishing, some other portions (see figure 35) have been completely ruined. These degenerate parts are situated either in extremely exposed positions at high altitudes or on wet moorland peat, both situations being highly unsuitable for tree planting.

The following is a list of plants compiled on the site of a portion of this decrepit plantation where the soil is wet, sour, peaty, and badly aërated:—

Copyright *W. B. Crump*

Figure 35.

Derelict Plantation.

Larches (*Larix decidua*), Beeches (*Fagus sylvatica*), etc., on wet, acidic peat. Ground vegetation of Heather (*Calluna vulgaris*), cross-leaved Heath (*Erica Tetralix*), purple Moor-grass (*Molinia caerulea*), Mat-grass (*Nardus stricta*), etc. Altitude 1600 feet (488 m.).

Sub-dominant

Erica Tetralix | Calluna vulgaris
Molinia caerulea | Eriophorum vaginatum

Abundant

Nardus stricta | Deschampsia flexuosa

Locally abundant

Empetrum nigrum | Vaccinium Myrtillus
Juncus squarrosus

Occasional

Potentilla erecta | Galium saxatile

This list was taken at an altitude of about 1500 feet (457 m.); and the plantation extends, or rather its remains extend, up to 1700 feet (518 m.). At altitudes higher than about 1550 feet (472 m.), however, the plantations of the district are, generally speaking, failures.

AFFORESTATION.

The question of the afforestation of waste lands in Britain has in recent years occupied the attention of the public; and this attention has recently been stimulated by the publication of a Government report.

As the present district comprises a large proportion of waste or uncultivated land, and as it contains numerous plantations, some successful and others unsuccessful, on parts of this waste land, a few remarks on the general subject are here given.

Much of the waste land of the district is utterly unfitted for immediate afforestation. This, in fact, applies to all peaty moorland which is dominated by such plants as the cotton-grasses (*Eriophorum vaginatum* or *E. angustifolium*), *Scirpus caespitosus*, heather (*Calluna vulgaris*), bilberry (*Vaccinium Myrtillus*), crowberry (*Empetrum nigrum*), and purple moor-grass (*Molinia caerulea*).

Before these sour and peaty places can be rendered fit for afforestation, a great deal of preliminary work is necessary;

and, until the peat, which should first be removed, can be profitably utilized in some way, the cost of the initial labour on such soils would be such as to render any plantations unprofitable from a financial point of view.

On the other hand, almost all the land which consists of calcareous grassland, and also much of the siliceous grassland dominated by the mat-grass (*Nardus stricta*), is fit, with a very small amount of preparatory labour, to be immediately put down to timber; and, if proper precautions be taken, there is no reason whatever why such plantations should not prove to be undertakings of a financially profitable nature.

However, the numerous derelict plantations on the Pennines, even on the grasslands, prove conclusively that reasonable precautions have frequently not been taken in the past; and this also applies not only to plantations laid down by private landowners but also to some recent attempts at afforestation on the part of municipal corporations. It is frequently overlooked that afforestation of uncultivated uplands is a very different matter from the laying down of plantations in lowland localities with a more genial climate; and this aspect of the case is one which does not appear to have been scientifically investigated by English foresters. Again, many of the unsuccessful plantations are of small size; and small plantations on exposed uplands cannot be expected to prosper. In a large plantation, the trees within the plantation receive shelter from those at the margin; but a small plantation is quickly devastated from end to end. Thirdly, the particular species of tree which is likely to flourish on the chosen site is frequently not sufficiently considered, although this would appear to be a matter of prime importance. One frequently finds in the decadent plantations at least a dozen species of trees and shrubs, some of which have never had any reasonable chance of reaching maturity; and it would appear that they have been obtained in an absurdly haphazard manner, from some lowland nurseryman. Other important precautions are often neglected; but enough has been said to indicate that, even on the more favourable sites, the afforestation of British uplands is a matter which must be undertaken in a more scientific spirit than has hitherto been the case if it has to have any reasonable probability of success.

Copyright *W. B. Crump*

Figure 36.

Reservoir among the moors.

Utilization of the Peat-Moors

It has been stated elsewhere (Moss, 1904) that the Pennine peat-moors represent a valuable English asset which is turned to little account. Grouse (*Lagopus scoticus*) are driven and shot over them, it is true; but considering the enormous rents paid by tenants for good grouse moors, it is surprising that more attention is not paid to the better cultivation of the heather and the bilberry, as these plants are much better adapted to the habits of the grouse than the cotton-grasses. By suitable encouragement, the former plants could be made to occupy much of the land now occupied by the latter.

Of late years, town and city corporations have utilized the peat-moors as gathering grounds for reservoirs (see figure 36); and thus an efficient water supply has been procured for the ever-growing manufacturing towns and villages which flank the Pennines.

Whilst the moors themselves are uninhabited, and have been so throughout the historic period, there is, as has often been shown (see Moss, 1904), abundant evidence to prove that neolithic man tenanted the sites of the present moorlands before the accumulation of the peat. The inhabitants of the moor-edges, up to a comparatively few years ago, possessed turf-cutting rights; but these, in nearly all cases, seem to have been lost. This is remarkable, as there is fuel enough in the Pennine peat to last the hill-side population for a thousand years. In addition to the value of the peat as fuel, the various products which might be manufactured from the peat could be made to furnish a satisfactory revenue, as is proved by the experience in certain foreign countries, such as Sweden. Finally, if the peat were gradually removed and utilized, the surface thus laid bare would, in many places, become fit for successful reclamation or afforestation.

APPENDIX I

SUMMARY AND RELATIONS OF THE PLANT COMMUNITIES OF THE PEAK DISTRICT

1. The Plant Formation of Calcareous Soils (Calcarion)

Group of Associations	Chief Associations	Subordinate Associations
Woods	Ash wood (Fraxinetum excelsioris)	
Scrub		Progressive scrub Retrogressive scrub
Grassland	Calcareous grassland (Festucetum ovinae)	Calcareous heath grassland Calcareous heath Vegetation of swamps Vegetation of "rakes"
Vegetation of screes and rocks		Vegetation of screes Vegetation of rocks
Limestone swamps		

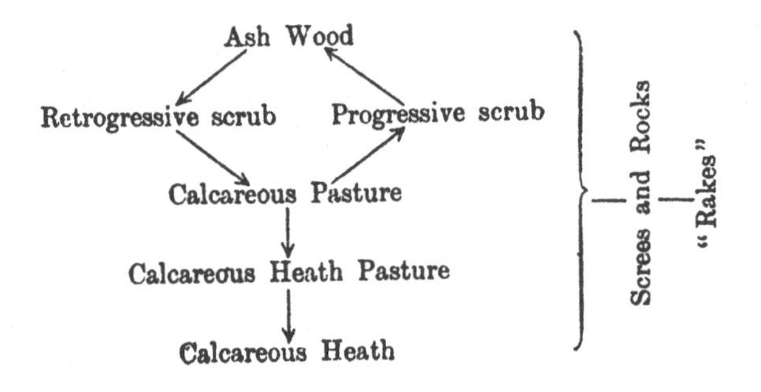

2 The Plant Formation of Siliceous Soils (Silicion)

Group of Associations	Chief Associations	Subordinate Associations
Woods	Birch wood (Betuletum tomentosae) Oak wood (Quercetum sessiliflorae)	
Scrub		Progressive scrub Retrogressive scrub
Grassland	Siliceous grassland (Nardetum strictae)	Siliceous grassland with much (a) Ulex Gallii (b) Pteris aquilina (c) Agrostis tenuis (d) Deschampsia flexuosa (e) Juncus effusus or J. effusus *forma* compactus (f) Calluna vulgaris
	Molinia grassland (Moliniëtum caeruleae)	Vegetation of swamps
Siliceous swamps		

3. The Plant Formation of the Acidic Peaty Soils (Oxodion)

Formation	Chief Associations	Subordinate Associations
Moor	Molinia moor (Moliniëtum caeruleae)	
	Heather moor (Callunetum vulgaris)	Heather moor with much (a) Vaccinium Myrtillus (b) Eriophorum vaginatum
	Bilberry moor (Vacciniëtum myrtilli)	
	Cotton-grass moor (Eriophoretum vaginati)	Cotton-grass moor with much Eriophorum angustifolium
		Retrogressive moor with much (a) Vaccinium Myrtillus (b) Empetrum nigrum (c) Rubus Chamaemorus (d) Bare peat

4. The Relationships of the Oxodion and the Silicion

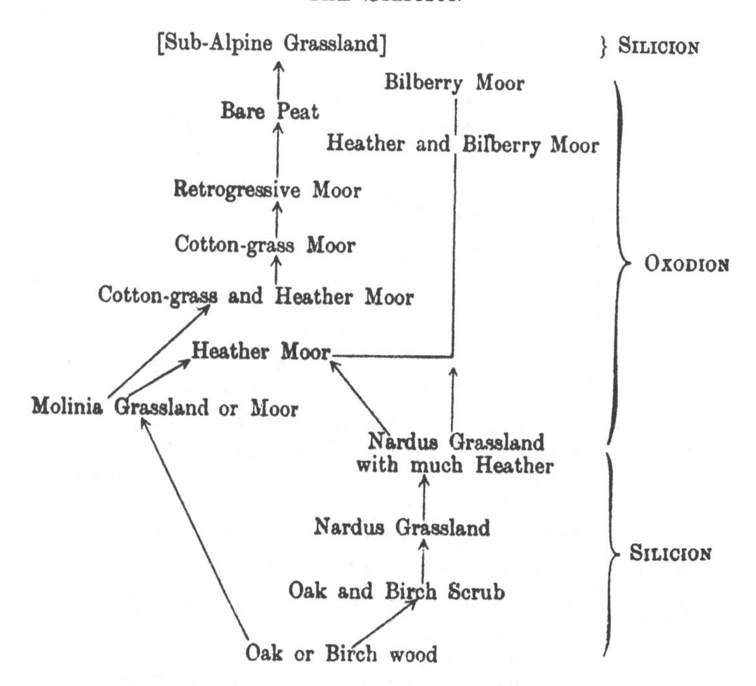

5. The Plant Formation of Fresh Waters

1. Associations of rapidly flowing non-calcareous waters.
2. Associations of rapidly flowing calcareous waters.
3. Associations of moving waters.
4. Associations of stagnant waters.
5. Associations of reed swamps.

APPENDIX II

SUMMARY OF BRITISH PLANT FORMATIONS AND ASSOCIATIONS

I. The Plant Formation of Fresh Waters.

- A. The Sub-formation of Foul Waters.
- B. The Sub-formation of nearly Stagnant Waters (*i.e.*, with no flood-currents).
  - 1. Associations of Submerged Plants (*e.g.*, Chareta).
  - 2. Associations of Plants with Floating Leaves (*e.g.*, Lemneta).
  - [2]3. Associations of Reed Swamps (*e.g.*, Phragmitidetum vulgaris).
- C. The Sub-formation of Slowly-moving Water (with periodical flood-currents and rich in dissolved mineral salts).
  - [2]1. Associations of Submerged Plants (*e.g.*, Ranunculetum circinati).
  - [2]2. Associations of Reed Swamps (*e.g.*, Glyceriëtum aquaticae).
- D. The Sub-formation of lake-margins, with well-aërated waters.
- E. The Sub-formation of Quickly-flowing Streams of hill and mountain slopes.
  - [1]1. Associations of Streams with Calcareous Waters.
  - [1]2. Associations of Streams with Non-calcareous Waters.
- F. The Sub-formation of Stagnant and Acidic Waters.

II. The Plant Formation of Salt and Brackish Waters.

- 1. Associations of Sea-weeds (*e.g.*, Laminariëtum digitatae).
- 2. Associations of submerged Marine Flowering Plants (*e.g.*, Zosteretum marinae).
- 3. Associations of Marine and Tidal Reed Swamps (*e.g.*, Spartineta).
- 4. Associations of Brackish Waters (*e.g.*, Ranunculetum baudotii).
- 5. Associations of Brackish Reed Swamps (*e.g.*, Scirpetum maritimi).

[1] Well represented in the Peak District.

[2] Fairly well represented in the Peak District.

III. The Plant Formation of Salt (NaCl) Soils.

1. Associations of open Salt Marshes (Salicorniëtum europaeae; S. ramosissimae).
2. Associations of intermediate or mixed Salt Marshes (*e.g.*, Staticetum maritimae).
3. Association of salt marsh grassland (*e.g.*, Glyceriëtum maritimae).
4. Associations of retrogressive Salt Marshes (*e.g.*, Atripli-cetum portulacoidis).
5. Associations of Spray-washed Rocks (*e.g.*, Crithmetum maritimi).
6. Associations of Strand Plants (*e.g.*, Atripliceta; Salsoletum kali). Transitional to dunes.
7. Associations of Maritime-fen Grassland. Transitional to fens.

IV. The Plant Formation of Sand Dunes and Shingle Banks.

1. Associations of Embryonic Dunes (*e.g.*, Agropyretum juncei).
2. Associations of Shifting Dunes (*e.g.*, Ammophiletum arenariae).
3. Associations of Fixed Dunes (*e.g.*, Festuceta).
4. Associations of Retrogressive Dunes.
5. Associations of Shingle Banks.

V. The Plant Formation of dry Sandy and Gravelly Soils.

1. Associations of dry woodlands.
   - *a.* Sub-association of *Fagus sylvatica*. Transitional to IX.
   - *b.* Sub-association of *Quercus* spp.
   - *c.* Sub-association of *Betula* spp.
   - *d.* Sub-association of *Pinus sylvestris.*
   - *e.* Mixed woods.
2. Associations of Scrub.
3. Associations of Sandy Grassland.

VI. The Plant Formation of Heaths.

1. Association of *Calluna vulgaris.*
   - *b.* Sub-association of *Erica cinerea.*
2. Associations of Heath Moors. Transitional to XII.

VII. The Plant Formation of the older Siliceous Soils.

[2]1. Association of Birch Woods (Betuletum pubescentis).
[1]2. Association of Oak Woods (Quercetum sessiliflorae).
[1]3. Associations of Scrub.
[1]4. Associations of Siliceous Grassland (*e.g.*, Nardetum strictae; Moliniëtum caeruleae). Moliniëta are transitional to moors.
[1]5. Associations of Swamps (*e.g.*, Juncetum effusi). Transitional to X.

VIII. The Plant Formation of Clayey Soils.

1. Association of damp Oak Woods (*e.g.*, Quercetum roboris).
   b. Sub-association of damp Oak-Hornbeam Woods.
2. Associations of Scrub.
3. Associations of Neutral Grassland.
4. Associations of Swamps. Transitional to X.

IX. The Plant Formation of Calcareous Soils.

1. Association of Beech Woods on Chalk (Fagetum sylvaticae).
2. Association of Yew Woods on Chalk (Taxetum baccatae).
[1]3. Association of Ash Woods (Fraxinetum excelsioris).
4. Association of Ash-Oak Woods on calcareous clays and marls.
[1]5. Associations of Scrub.
[1]6. Associations of Calcareous Grassland (*e.g.*, Festucetum ovinae)
[1]7. Associations of Swamps. Transitional to X.

X. The Plant Formation of Marshy Soils (*i.e.*, soils subject to periodical inundations).

1. Associations of Marsh Woods (*e.g.*, Alneta, Saliceta).
2. Associations of Marsh Scrub.
3. Associations of Marsh Grassland.

XI. The Plant Formation of Peaty Soils with alkaline Waters (=the Fen Formation).

1. Associations of Fens.
2. Associations of Fen Scrub.
3. Associations of Fen Woods.
4. Associations of Fen Grassland. Transitional to X.

XII. The Plant Formation of Peaty Soils with acidic Waters (=the Moor Formation).

1. Associations of Bog-mosses (Sphagneta).
2. Associations of Moor Scrub.
3. Associations of Moor Woods, *e.g.* (Pineta, Betuleta).
4. Association of *Rhyncospora alba*. Transitional to XI.
5. Association of *Eriophorum angustifolium*.
6. Association of *Scirpus caespitosus*.
7. Retrogressive Associations with *Rhacomitrium lanuginosum*.
[1]8. Association of Cotton-grass Moor (Eriophoretum vaginati).
[1]9. Retrogressive Associations with *Vaccinium Myrtillus*, *Empetrum nigrum*, and *Rubus Chamaemorus*.
[1]10. Association of Bilberry Moor (Vacciniëtum myrtilli).
   [1]*b.* Sub-association of *Vaccinium Vitis-idaea*.
[1]11. Association of Heather Moor (Callunetum vulgaris).
[2]12. Association of Grass Moor. Transitional to VII.

XIII. The Plant Formation of Alpine Summits.

1. Associations of Calcareous rocks.
2. Associations of Non-calcareous rocks.

XIV. The Plant Formation of Cultivated Land.

1. Associations of the Alluvial Zone.
   a. Permanent Pasture.
   b. Arable Land.

[2]2. Associations of the Wheat Zone.
   [2]*a.* Permanent Pasture.
   [2]*b.* Arable Land.
   c. Orchards.
   d. Market Gardens.

[1]3. Associations of the No-wheat Zone.
   [1]*a.* Permanent Pasture.
   [2]*b.* Arable Land.

APPENDIX III

LIST OF WORKS REFERRED TO IN THE TEXT

ACKROYD, W. (1899). "On Halifax Waters"; in *The Halifax Naturalist*, III. pp. 120–1; Halifax.

ARNOLD-BEMROSE, H. H. (1907). "The Toadstones of Derbyshire"; in *Quart. Journ. Geol. Soc.*, LXIII. pp. 241–281; London.

BAGNALL, J. E. (1901). "The Flora of Staffordshire"; in *Journ. of Bot.*, XXXIX. supp.; London. Also published separately; London.

BARROW, G. (1903). "The Geology of the Cheadle Coal Field"; London. (Mem. Geol. Survey, England and Wales.)

BARTHOLOMEW, J. G. AND HERBERTSON, A. J. (1899). "Physical Atlas, Part III. "Meteorology"; Edinburgh.

BENNETT, A. (1905). See also Watson, H. C.

—— (1908 *a*). "*Potamogeton pennsylvanicus* in England"; in *The Naturalist*, No. 612, pp. 10–11; Hull.

—— (1908 *b*). "The Halifax Potamogeton"; in *The Naturalist*, No. 621, pp. 373–5; Hull.

BLYTT, A. (1882). Die Theorie der wechselnden Kontinentalen und insularen Klimate; in Engler's *Bot. Jahrb.*, II.

—— (1893). Zur Geschichte der nordeuropäischen, besonders der norwegischen Flora; *ib.*

British Rainfall (Annual periodical, edited by H. R. Mill), 1866–1911; London.

BROCKMANN-JEROSCH, H. (1907). "Die Flora des Puschlav (Bezirk Bernina, Kanton Graubünden) und ihre Pflanzengesellschaften"; Leipzig.

BUCHAN, A. (1862). "The Meteorological Conditions which Determine the Profitable and Unprofitable Culture of Farm Crops in Scotland"; in *Quart. Rep. Meteorol. Soc. of Scotland*, pp. 2–12; Edinburgh.

BURRELL, B. A. (1900). "The Composition of Some Malham Waters"; in *Proc. Yorks. Geol. Soc.*, XIV. pp. 45–8.

CAJANDER, A. K. (1903). "Beiträge zur Kenntnis der Vegetation der Alluvionen des nördlichen Eurasiens, I. Die Alluvionen des unteren Lena-Thales"; in *Act. Soc. Sc. Fenn.* XXXII.; Helsingfors.

—— (1904). "Ein Beitrag zur Entwickelungsgeschichte der nordfinnischen Moore"; in *Fennia*, 20, 6; Helsingfors.

CLEMENTS, F. E. (1904). "The Development and Structure of Vegetation"; Lincoln, Neb., U.S.A.
—— (1905). "Research Methods in Ecology"; Lincoln, Neb., U.S.A.
—— (1907). "Plant Physiology and Ecology"; London.
COHEN, J. B. (1900). "Air of Towns"; in *Rep. Brit. Ass.* (Bradford); London.
—— AND RUSHTON, A. G. (1909). "The Nature and Effect of Air Pollution by Smoke"; see *Nature*, LXXXI. pp. 468–9; London.
CONWENTZ, H. (1910). "The Care of Natural Monuments"; Cambridge.
COWLES, H. C. (1911). "The Causes of Vegetative Cycles"; in *Bot. Gaz.*, LI. pp. 161–183; Chicago.
CRAMPTON, C. B. (1911). "The Vegetation of Caithness considered in Relation to the Geology"; (privately printed and published).
CROSSLAND, C. (1904). See Crump and Crossland.
CRUMP, W. B. AND CROSSLAND, C. (1904). "The Flora of Halifax"; Halifax. (First published serially in *The Halifax Naturalist*, 1896–1904. Flowering plants and Pteridophytes by Crump: mosses, liverworts, Algae, Fungi, and lichens by Crossland.)
DAKINS, J. R. (1869, 1887). See Green, Foster, and Dakins.
DALE, E. (1900). "The Scenery and Geology of the Peak of Derbyshire"; London and Buxton.
DAVEY, F. H. (1909). "Flora of Cornwall" (flowering plants and ferns); Penryn.
[DEFOE, D.] (1724–5). "A Tour thro' the whole Island of Great Britain," by "A Gentleman"; London.
DRUDE, O. (1896). "Deutschlands Pflanzengeographie, I."; Stuttgart.
ELWES, H. J. AND HENRY, A. (1905–1913). "The Trees of Great Britain and Ireland"; Edinburgh. (Seven volumes.)
ENGLER, A. (1907). "Syllabus der Pflanzenfamilien," 5th ed.; Berlin.
ERNST, A. (1908). "The New Flora of the Volcanic Island of Krakatau"; Cambridge. English translation by A. C. Seward.
FAREY, J. (1811–3). "General view of the Agriculture and Minerals of Derbyshire"; London.
FERNALD, M. L. (1908). "Notes on *Potamogeton pennsylvanicus* Cham."; in *The Naturalist*, No. 621, pp. 375–6; Hull.
FISHER, W. R. See Schimper.
FLAHAULT, C. (1897). "Essai d'une Carte Botanique et Forestière de la France"; in *Ann. de Geogr.*, V. or VI. pp. 289–312 (with coloured vegetation map); Paris.
—— (1901). "La Flore et la Vegetation de la France," pp. 1–52; in *Flore descriptive et illustrée de la France* by the abbé H. Coste; Paris.
—— AND SCHRÖTER, C. (1910). "Phytogeographical Nomenclature: Reports and Propositions": also "Nomenclature Phytogéographique: Votes et Remarques": IIIe Congrès international de Botanique, Bruxelles, May, 1910; Zurich.
FOSTER, C. LE N. (1869, 1887). See Green, Foster, and Dakins.
FRÜH, J. AND SCHRÖTER, C. (1904). "Die Moore der Schweiz"; Bern.
GEIKIE, J. (1906). "From the Ice Age to the Present"; in *The Scott. Geogr. Mag.* XXII. pp. 397–407; Edinburgh.

Gèze, J. B. (1908). "Notes d'éphaphisme chemique"; in *Bull. Soc. Bot France*, LV. pp. 462–466; Paris.

Gradmann, R. (1909). "Über Begriffsbildung in der Lehre von Pflanzenformationen"; in Engler's *Botan. Jahrb.* XXXVIII. Beibl. 99.

Graebner, P. (1901). "Die Heide Norddeutschlands"; in Engler u. Drude, *Veg. d. Erde*, v.; Leipzig. (A review with special reference to work in Britain, by W. G. Smith, in *Scott. Geogr. Mag.*, Nov. 1902; Edinburgh.)

—— (1902). See also Warming.

—— (1909). Die Pflanzenwelt Deutschlands; Leipsig.

Green, A. H., Foster, C. le N. and Dakins, J. R. (1869); 2nd ed. 1887. "The Geology of the Carboniferous Limestone, Yoredale Rocks, and Millstone Grit of north Derbyshire"; London. (Mem. Geol. Surv., England and Wales.)

Grisebach, H. R. A. (1846). "Report on Botanical Geography during the year 1842"; "Report on Botanical Geography during the year 1843"; in *Reports and Papers on Botany*, pp. 57–212; London (Ray Soc.).

—— (1849). "Report on the Progress of Geographical Botany during the year 1844"; "Report on the Progress of Geographical and Systematic Botany during the year 1845"; in *Reports and Papers on Botany*, pp. 317–493; London (Ray Soc.).

Hall, A. J. (1908). "The Soil"; London.

Hall, A. J. and Russell, E. J. (1911). "A Report on the Agriculture and Soils of Kent, Surrey, and Sussex"; London.

Hardy, M. (1902). "Botanical Geography and the Biological Utilization of the Soil"; in *Scott. Geogr. Mag.* XVIII.

—— (1905). "Esquisse de la géographie et de la végétation des Highlands D'Ecosse"; Paris. (See also "La Végétation des Highlands D'Ecosse"; in *Ann. de Géogr.* XV, 1905; Paris. See also "Botanical Survey of Scotland. A General Map of the Highlands with a sketch of the History of the Methods"; in *Scott. Geogr. Mag.* XXII. pp. 229–241, with coloured vegetation map, 1906; Edinburgh.)

—— (1909). See *Report Brit. Assoc.*

Hedgcock, G. G. (1902). "The Relation of the Water Content of the soil to certain Plants, principally Mesophytes"; Lincoln, Neb., U.S.A.

Henry, A. See Elwes and Henry.

Herbertson, A. J. See Bartholomew and Herbertson.

Herbertson, A. J. (1910). "Geography and some of its Present Needs"; in *Report Brit. Assoc. for* 1910, pp. 640–649 (1911).

Hind, Wheelton (1897). "The Yoredale Series"; in *Geol. Mag.* 205–213.

—— and Howe, J. A. (1901). "The Geological Succession and Paleontology of the Beds between the Millstone Grit and the Limestone-Massif at Pendle Hill and their Equivalents in certain other parts of Britain"; in *Quart. Journ. Geol. Soc.* LVII. pp. 347–404.

Holland, A. (1808). "The Agriculture of Cheshire"; London.

Hooker, J. D., Sir (1884). "The Student's Flora of the British Islands" (3rd ed.); London.

HORRELL, E. C. (1900). "The European Sphagnaceae (after Warnstorf)"; in *Journ. of Bot.* XXXVIII., April to Dec.; London. Also published separately; London.

KIRCHNER, O. See Schröter und Kirchner.

KNOBLAUCH, E. See Warming.

KRASSNOFF, A. (1886). "Geobotanical Researches in the Kalmuk Steppe." See Engler's *Bot. Jahrb.* (Litteraturbericht) x. 1889.

KRAUS, G. (1911). "Boden and Klima"; Jena.

KRAUSE, E. H. L. (1892). "Die Heide"; in Engler's *Bot. Jahrb.* XIV.

LAMPLUGH, G. W. (1906). "On British Drifts and the Interglacial Problem"; in *Rep. Brit. Assoc.* (York), pp. 532–558; London.

LEES, F. A. (1888). "The Flora of West Yorkshire: with a sketch of the climatology and lithology" (Phanerogams and Cryptogams); London. (New supplement in preparation.)

LEWIS, F. J. (1904 *a*). "Geographical Distribution of Vegetation of the Basins of the rivers Eden, Tees, Wear, and Tyne, Part I"; in *Geogr. Journ.* XXIII. pp. 313–331 (with coloured vegetation map); London.

—— (1904 *b*). "Geographical Distribution of Vegetation of the Basins of the rivers Eden, Tees, Wear, and Tyne, Part II"; in *Geogr. Journ.* XXIV. pp. 267–285 (with coloured vegetation map); London.

LEWIS, F. J. (1905–7). "The Plant Remains in the Scottish Peat Mosses"; in *Trans. Roy. Soc. Edinburgh*; Part I, XLI. (1905), pp. 699–723; Part II (1906), pp. 335–360; Part III (1907), pp. 33–70; Edinburgh.

LINTON, W. R. (1903). "Flora of Derbyshire: flowering plants, higher Cryptogams, mosses and Hepatics, Characeae"; London.

MARGERISON, S. (1907–9). "The Vegetation of some Disused Quarries"; in *Bradford Scientific Journal*, 1907–9; Bradford. Also published separately; Bradford.

MASSART, J. (1910). "Esquisse de la Géographie Botanique de la Belgique," in *Rec. de l'Institut Léo Errera*, t. supp. VII. bis; Bruxelles.

MILL, H. R. See *British Rainfall.*

MOORE, S. See Lord de Tabley.

MOSS, C. E. (1903). See Smith and Moss.

—— (1904). "Peat Moors of the Pennines: their Age, Origin and Utilization"; in *Geogr. Journ.* XXIII. pp. 660–671; London.

—— (1907 *a*). "Geographical Distribution of Vegetation in Somerset: Bath and Bridgwater District" (with coloured vegetation map); Roy. Geogr. Soc. (also Edward Stanford), London.

—— (1907 *b*). "Xerophily and the Deciduous Habit"; in *New Phyt.* VI. pp. 183–185.

—— (1910 *a*). "British Oaks"; in *Journ. of Bot.*, XLVIII. pp. 1–8 and 33–39; London.

—— (1910 *b*). "The Fundamental Units of Vegetation: Historical Development of the Concepts of the Plant Association and the Plant Formation"; in *New Phyt.* IX. pp. 18–53; Cambridge. Also published separately; Cambridge.

—— (1911). "Plant Ecology"; in *Encycl. Brit.* ed. XI., vol. IX. pp. 113–149; London.

Moss, E. C., Rankin, W. M. and Tansley, A. G. (1910). "The Woodlands of England"; in *New Phyt.* Also published separately; Cambridge.

Murray, H. See Weiss and Murray.

Nägeli, C. von (1865). "Ueber die Bedingungen des Vorkommens von Arten und Varietäten innerhalb ihres Verbreitungsbezirkes." Sitzungsberichte der bayrischen Akadamie; Bd. I.

Naturalist, The; Hull.

New Phytologist, The; Cambridge.

Nilsson, A. (1902). "Zur Ernährungsoekonomie der Pflanzen"; in *Helsingfors Centraltryckeri*, pp. 1–3; Helsingfors.

Oliver, F. W. (1893). "On the Effects of Urban Fog upon Cultivated Plants"; in *Journal of the Roy. Horticultural Soc.* Part I, XVI. pp. 1–59; London.

Ostenfeld, C. H. (1901). "Geology"; pp. 24–31, in *Botany of the Faröes*, I.; Copenhagen, Christiana, and London.

—— (1908). "The Land Vegetation of the Faröes," pp. 867–1026, in *Botany of the Faröes*, III.; Copenhagen, Christiana, and London Also published separately, Copenhagen.

Öttli, M. (1905). "Beiträge zur Ökologie der Felsflora"; Zurich.

Painter, W. H. (1889). "A Contribution to the Flora of Derbyshire"; London and Derby.

—— (1899). "Notes supplementary to the Flora of Derbyshire"; "List of Derbyshire Mosses"; in *The Naturalist.*

Pethybridge, G. H. and Praeger, R. Ll. (1905). "The Vegetation of the District lying south of Dublin"; in *Proc. Roy. Irish Acad.* XXV. B, no. 6, pp. 124–180 (with coloured vegetation map); Dublin.

Praeger, R. Ll. (1905). See Pethybridge and Praeger.

—— (1909). "A Tourist's Flora of the West of Ireland"; Dublin.

Ramann, E. (1905). "Bodenkunde," 2nd ed.; Berlin.

Rankin, W. M. (1903). See Smith and Rankin.

—— (1909). See Kendall, Dean and Rankin.

—— (1910). See Moss, Rankin and Tansley.

—— (1910). "The Peat Moors of Lonsdale: an Introduction"; in *The Naturalist*, no. 638 pp. 119–122, and no. 639, pp. 153–161; Hull.

Rübel, E. (1911). "Pflanzengeographische Monographie des Berninagebietes" (with coloured vegetation map); in Engler's *Botan. Jahrb.*; Leipzig.

Rushton, A. G. (1909). See Cohen and Rushton.

Russell, E. J. See Hall and Russell.

Samuelsson, G. (1910). "Scottish Peat Mosses"; in *B. Geol. I. Univ. Upsala*, pp. 197–260.

Scarth, G. W. (1911). "The Grassland of Orkney"; in *Trans. and Proc. Bot. Soc. Edinb.* XXIV. pp. 143–163.

Schimper, A. F. W. (1898). "Pflanzengeographie auf physiologischer Grundlage"; Jena. English tr. by Fisher, "Plant Geography upon a Physiological Basis," 1903–4; Oxford.

Schröter, C. (1904). See Früh and Schröter.

SCHRÖTER, C. (1910). See Flahault and Schröter.

—— und KIRCHNER, O. (1896–1902). "Die Vegetation des Bodensees"; Lindau.

SCHUSTER, A. *Report on the Investigation of the Upper Atmosphere... Glossop*, 1908–9. (Private Monthly Periodical; Manchester.)

SCHOUW, J. F. (1822). "Grundtvaek til en almindelig Plantegeografie"; Kjobenhavn. German tr. "Grundzüge einer allegemeinen Pflanzengeographie," 1823; Berlin.

SENDTER, O. (1860). Die Vegetationsverhältnisse des Bayerischen Waldes; München.

SEWARD, A. C. See Ernst.

SIBLEY, T. F. (1908). "The Faunal Succession in the Carboniferous Limestone (Upper Avonian) of the Midland Area (North Derbyshire and North Staffordshire)"; in *Quart. Journ. Geol. Soc.* LXIV. pp. 34–82; London.

SMITH, R. (1900*a*). "Botanical Survey of Scotland: I, Edinburgh District" (with coloured vegetation map) in *Scott.*; *Geogr. Mag.* XVI. pp. 385–416; Edinburgh. Also Bartholomew, Edinburgh.

—— (1900 *b*). "Botanical Survey of Scotland: II, North Perthshire District" (with coloured vegetation map); in *Scott. Geogr. Mag.* XVI. pp. 441–467; Edinburgh. Also Bartholomew, Edinburgh.

—— and SMITH, W. G. (1904–5). "Botanical Survey of Scotland: III and IV, Forfar and Fife" (with two coloured vegetation maps); in *Scott. Geogr. Mag.* XX. pp. 617–628, XXXI. pp. 4–23, 57–83, and 117–126; Edinburgh.

SMITH, W. G. and MOSS, C. E. (1903). "Geographical Distribution of Vegetation in Yorkshire: Part I, Leeds and Halifax District" (with coloured vegetation map); in *Geogr. Journ.* XXI. pp. 375–401; London. Also Bartholomew, Edinburgh.

—— and RANKIN, W. M. (1903). "Geographical Distribution of Vegetation in Yorkshire: Part II, Harrogate and Skipton District" (with coloured vegetation map); in *Geogr. Journ.* XXII. pp. 149–178; London. Also Bartholomew, Edinburgh.

—— (1911). "The Vegetation of Woodlands"; in *Trans. of Roy. Scott. Arboricultural Soc.*

——. See also Gräbner.

STEBLER, F. G. (1906). Der Kalkgehalt einger Esparsetteböden; in *Landu. Jahrb. der Schweiz.*

TABLEY, Lord DE (1899). "The Flora of Cheshire" (flowering plants and Pteridophytes); London. Edited by Spencer Moore.

TANSLEY, A. G. (1910). See Moss, Rankin and Tansley.

—— (1911). "Types of British Vegetation"; Cambridge University Press.

THURMANN, J. (1849). "Essai de phytostatique, appliqué à la chaine du Jura et aux contrées voisins"; Bern.

VAHL, M. See Warming.

WARMING, E. (1895). "Plantesamfund"; Copenhagen. German tr. by Knoblauch, 1896; Berlin. 2nd German ed. by Graebner, 1902; Berlin.

WARMING, E., assisted by VAHL, M. (1909). "Ecology of Plants: an introduction to the study of plant communities"; Oxford.

WARNSTORF, C. See Horrell.

WARREN, Hon. J. BYRNE LEICESTER. See Lord de Tabley.

WATSON, H. C. (1832). "Outlines of the Geographical Distribution of British Plants"; Edinburgh.

—— (1835). "Remarks on the Geographical Distribution of British Plants"; London.

—— (1836). "Observations on the Construction of Maps for illustrating the Geographical Distribution of Plants"; in *Mag. Nat. Hist.* IX.

—— (1843). "The Geographical Distribution of British Plants"; London.

—— (1847–1859). "Cybele Britannica," 4 vols.; London. (Also Supp. I, 1860; London.)

—— (1868–1870). "A Compendium of the Cyb. Brit."; Thames Ditton. (Also Supp. 1872; Thames Ditton.)

—— (1873–4). "Topographical Botany"; Thames Ditton.

—— (1883). "Topographical Botany," 2nd ed.; London.

WATSON, H. C. [(1905). "Supplement to Topographical Botany, ed. 2"; in *Journ. of Bot.* XLIII. supp., by Arthur Bennett. Also published separately; London.]

WATSON, W. (1909). "The Distribution of Bryophytes in the Woodlands of Somerset"; in *New Phyt.*, VIII. pp. 90–96.

WEBER, C. A. (1908). "Die wichtigsten Humus- und Torfarten und ihre Beteiligung an dem Aufbau norddeutscher Moore"; Berlin.

WEISS, F. E. (1908). "The Dispersal of Fruits and Seeds by Ants"; in *The New Phyt.* VII. pp. 23–28; Cambridge.

—— (1909). "The Dispersal of the Seeds of the Gorse by Ants"; in *The New Phyt.* VIII. pp. 81–89; Cambridge.

WEISS, F. E. and MURRAY, H. (1909). "On the Occurrence and Distribution of some Alien Aquatic Plants in the Reddish Canal"; in *Mem. Proc. Manchester Lit. and Phil. Soc.* LIII. ii.; Manchester.

WHELDON, J. A. and WILSON, A. (1907). "The Flora of West Lancashire" (flowering plants, Pteridophytes, mosses, Hepatics and lichens); Eastbourne.

WILLIAMS, F. N. (1900 to 1910). "Prodromus Florae Britannicae, vol. I; Brentford.

WILSON, A. (1900). "The Great Smoke-Cloud of the North of England and its Influence on Plants"; in *Rep. Brit. Assoc.* pp. 930–1 (Bradford); London.

—— (1907). See also Wheldon and Wilson.

WOODHEAD, T. W. (1904). "Notes on the Bluebell"; in *The Naturalist*, no. 565, pp. 41–48, and no. 566, pp. 81–88.

—— (1906). "Ecology of Woodland Plants in the Neighbourhood of Huddersfield"; in *Linn. Soc. Journ.*, Bot., pp. 333–406; London.

YAPP, R. H. (1908). "Wicken Fen"; in *New Phyt.* VII.

—— (1909). "On Stratification in the Vegetation of a Marsh, and its Relation to Evaporation and Temperature"; in *Ann. of Botany*, XXIII. pp. 275–320.

INDEX

Printed in the United States
By Bookmasters